T0302291

Nuclear Radiation
Detection Materials

MATERIALS RESEARCH SOCIETY
SYMPOSIUM PROCEEDINGS VOLUME 1038

Nuclear Radiation Detection Materials

Symposium held November 28–29, 2007, Boston, Massachusetts, U.S.A.

EDITORS:

Arnold Burger
Fisk University
Nashville, Tennessee, U.S.A.

Larry Franks
Special Technologies Laboratory
Santa Barbara, California, U.S.A.

Dale L. Perry
Lawrence Berkeley National Laboratory
Berkeley, California, U.S.A.

Michael Schieber
The Hebrew University of Jerusalem
Jerusalem, Israel

Materials Research Society
Warrendale, Pennsylvania

CAMBRIDGE
UNIVERSITY PRESS

University Printing House, Cambridge CB2 8BS, United Kingdom

One Liberty Plaza, 20th Floor, New York, NY 10006, USA

477 Williamstown Road, Port Melbourne, VIC 3207, Australia

314-321, 3rd Floor, Plot 3, Splendor Forum, Jasola District Centre, New Delhi - 110025, India

79 Anson Road, #06-04/06, Singapore 079906

Cambridge University Press is part of the University of Cambridge.

It furthers the University's mission by disseminating knowledge in the pursuit of education, learning and research at the highest international levels of excellence.

www.cambridge.org
Information on this title: www.cambridge.org/9781558999855

Single article reprints from this publication are available through University Microfilms Inc., 300 North Zeeb Road, Ann Arbor, MI 48106

CODEN: MRSPDH

Published by:
Materials Research Society
506 Keystone Drive
Warrendale, PA 15086
Telephone (724) 779-3003
Fax (724) 779-8313
Web site: http://www.mrs.org/

A catalogue record for this publication is available from the British Library

ISBN 978-1-558-99985-5 Hardback

Cambridge University Press has no responsibility for the persistence or accuracy of URLs for external or third-party internet websites referred to in this publication, and does not guarantee that any content on such websites is, or will remain, accurate or appropriate.

CONTENTS

*Invited Paper

*Invited Paper

PREFACE

This volume is a collection of papers presented at Symposium O, "Nuclear Radiation Detection Materials," held November 28–29 at the 2007 MRS Fall Meeting in Boston, Massachusetts. The purpose of this symposium was to bring together the materials science community with their characterization techniques and the community that grows nuclear radiation detection materials and studies their performance. The broad range of materials and characterization methods presented at this symposium indicates that these goals were achieved.

This symposium provided a venue for the presentation of the latest results and discussion of radiation detection materials from both experimental and theoretical standpoints. As advances are made in this area of materials, additional experimental and theoretical approaches are being used to both guide the growth of materials and to characterize the materials that have a wide array of applications for detecting different types of nuclear radiation. The scope of the types of detector materials for semiconductors and scintillators included a wide variety of molecular compounds such as cadmium zinc telluride (CZT), lanthanum halides, and others. An additional class of scintillators included those based on organic compounds and glasses. Ideally, desired materials used for radiation detection have attributes such as appropriate-range bandgaps, high atomic numbers of the central element, high densities, performance at room temperature, and strong mechanical properties, and are low cost in terms of their production. There are significant gaps in the knowledge related to these materials in these areas that are very important in making radiation detection materials that are higher quality in terms of their reproducible purity, homogeneity, and mechanical integrity. The topics that were the focal point of this symposium addressed these issues so that much better detectors may be made in the future.

Several of the topics included in the scope of areas of interest in this symposium were characterizational experimental results such as surface and bulk effects, interfacial phenomena such as contacting and contact adhesion, crystallographic polarity, and surface passivation. Physical and mechanical properties included interestelectric charge compensation mechanisms, charge collection, thermal transport, hardness, and plasticity.

Arnold Burger
Larry Franks
Dale L. Perry
Michael Scheiber

August 2008

MATERIALS RESEARCH SOCIETY SYMPOSIUM PROCEEDINGS

Volume 1024E —Combinatorial Methods for High-Throughput Materials Science, D.S. Ginley, M.J. Fasolka,
A. Ludwig, M. Lippmaa, 2008, ISBN 978-1-60511-000-4
Volume 1025E —Nanoscale Phenomena in Functional Materials by Scanning Probe Microscopy, L. Degertekin,
2008, ISBN 978-1-60511-001-1
Volume 1026E —Quantitative Electron Microscopy for Materials Science, E. Snoeck, R. Dunin-Borkowski,
J. Verbeeck, U. Dahmen, 2008, ISBN 978-1-60511-002-8
Volume 1027E —Materials in Transition—Insights from Synchrotron and Neutron Sources, C. Thompson,
H.A. Dürr, M.F. Toney, D.Y. Noh, 2008, ISBN 978-1-60511-003-5
Volume 1029E —Interfaces in Organic and Molecular Electronics III, K.L. Kavanagh, 2008,
ISBN 978-1-60511-005-9
Volume 1030E —Large-Area Processing and Patterning for Active Optical and Electronic Devices, V. Bulović,
S. Coe-Sullivan, I.J. Kymissis, J. Rogers, M. Shtein, T. Someya, 2008, ISBN 978-1-60511-006-6
Volume 1031E —Nanostructured Solar Cells, A. Luque, A. Marti, 2008, ISBN 978-1-60511-007-3
Volume 1032E —Nanoscale Magnetic Materials and Applications, J-P. Wang, 2008, ISBN 978-1-60511-008-0
Volume 1033E —Spin-Injection and Spin-Transfer Devices, R. Allenspach, C.H. Back, B. Heinrich, 2008,
ISBN 978-1-60511-009-7
Volume 1034E —Ferroelectrics, Multiferroics, and Magnetoelectrics, J.F. Scott, V. Gopalan, M. Okuyama, M. Bibes,
2008, ISBN 978-1-60511-010-3
Volume 1035E —Zinc Oxide and Related Materials—2007, D.P. Norton, C. Jagadish, I. Buyanova, G-C. Yi, 2008,
ISBN 978-1-60511-011-0
Volume 1036E —Materials and Hyperintegration Challenges in Next-Generation Interconnect Technology,
R. Geer, J.D. Meindl, R. Baskaran, P.M. Ajayan, E. Zschech, 2008, ISBN 978-1-60511-012-7
Volume 1037E —Materials, Integration, and Technology for Monolithic Instruments II, D. LaVan, 2008,
ISBN 978-1-60511-013-4
Volume 1038— Nuclear Radiation Detection Materials, D.L. Perry, A. Burger, L. Franks, M. Schieber, 2008,
ISBN 978-1-55899-985-5
Volume 1039— Diamond Electronics—Fundamentals to Applications II, R.B. Jackman, C. Nebel, R.J. Nemanich,
M. Nesladek, 2008, ISBN 978-1-55899-986-2
Volume 1040E —Nitrides and Related Bulk Materials, R. Kniep, F.J. DiSalvo, R. Riedel, Z. Fisk, Y. Sugahara,
2008, ISBN 978-1-60511-014-1
Volume 1041E —Life-Cycle Analysis for New Energy Conversion and Storage Systems, V.M. Fthenakis,
A.C. Dillon, N. Savage, 2008, ISBN 978-1-60511-015-8
Volume 1042E —Materials and Technology for Hydrogen Storage, G-A. Nazri, C. Ping, A. Rougier,
A. Hosseinmardi, 2008, ISBN 978-1-60511-016-5
Volume 1043E —Materials Innovations for Next-Generation Nuclear Energy, R. Devanathan, R.W. Grimes,
K. Yasuda, B.P. Uberuaga, C. Meis, 2008, ISBN 978-1-60511-017-2
Volume 1044 — Thermoelectric Power Generation, T.P. Hogan, J. Yang, R. Funahashi, T. Tritt, 2008,
ISBN 978-1-55899-987-9
Volume 1045E —Materials Science of Water Purification—2007, J. Georgiadis, R.T. Cygan,
M.M. Fidalgo de Cortalezzi, T.M. Mayer, 2008, ISBN 978-1-60511-018-9

MATERIALS RESEARCH SOCIETY SYMPOSIUM PROCEEDINGS

Prior Materials Research Society Symposium Proceedings available by contacting Materials Research Society

Mater. Res. Soc. Symp. Proc. Vol. 1038 © 2008 Materials Research Society   1038-O01-03

Basic Materials Studies of Lanthanide Halide Scintillators

F. P. Doty[1], Douglas McGregor[2], Mark Harrison[1,2], Kip Findley[3], Raulf Polichar[4], and Pin Yang[5]

[1]Engineered Materials Dept., Sandia National Labs, Livermore, CA, 94550
[2]Dept. of Mechanical and Nuclear Engineering, Kansas State University, Manhattan, KS, 66506
[3]School of Mechanical and Materials Engineering, Washington State University, Pullman, WA, 99164
[4]SAIC, San Diego, CA, 92127
[5]Ceramic and Glass Dept., Sandia National Labs, Albuquerque, NM, 87185

ABSTRACT
Cerium and lanthanum tribromides and trichlorides form isomorphous alloys with the hexagonal UCl3 type structure, and have been shown to exhibit high luminosity and proportional response, making them attractive alternatives for room temperature gamma ray spectroscopy. However the fundamental physical and chemical properties of this system introduce challenges for material processing, scale-up, and detector fabrication. In particular, low fracture stress and perfect cleavage along prismatic planes cause profuse cracking during and after crystal growth, impeding efforts to scale this system for production of low cost, large diameter spectrometers. We have reported progress on basic materials science of the lanthanide halides. Studies to date have included thermomechanical and thermogravimetric analyses, hygroscopicity, yield strength, and fracture toughness. The observed mechanical properties pose challenging problems for material production and post processing; therefore, understanding mechanical behavior is key to fabricating large single crystals, and engineering of robust detectors and systems. Analysis of the symmetry and crystal structure of this system, including identification of densely-packed and electrically neutral planes with slip and cleavage, and comparison of relative formation and propagation energies for proposed slip systems, suggest possible mechanisms for deformation and crack initiation under stress. The low c/a ratio and low symmetry relative to traditional scintillators indicate limited and highly anisotropic plasticity cause redistribution of residual process stress to cleavage planes, initiating fracture. Ongoing work to develop fracture resistant lanthanide halides is presented.

INTRODUCTION
Lanthanum halide scintillators have enabled scintillating gamma ray spectrometers competitive with room temperature semiconductors[1,2], providing similar energy resolution with larger active volumes than available CdZnTe detectors, making such applications as hand held radioisotope spectrometers practical [3]. However, increasing the active volume to larger sizes needed for applications in nuclear nonproliferation and homeland security has proven difficult due to profuse cracking during crystal growth and subsequent processing. Therefore basic studies of the materials science of the lanthanide halide system are needed to determine the causes of cracking, and develop strategies to scale the crystal growth.

RESULTS AND DISCUSSION
Cleavage and slip
Results of materials property measurements made by Sandia National Laboratories and collaborators were recently reviewed[4],[5] and a strategy to strengthen these materials was proposed[6]. The key results were determination of the high degree of anisotropy of thermal expansion, plasticity, and fracture inherent to the crystal structure. In particular, the materials fail mechanically by brittle fracture, usually along very well defined cleavage planes

This anisotropy can be visualized in reference to the figure below, which represents adjacent lanthanide ion positions (green) in relation to some of the halogen ion positions (red). (Note that the polygons are marked by positions anion centers; the anion size has been distorted.) The rectangular faces of the prisms are parallel to the c-axis, and those faces adjoining the prisms represent the 11.0 planes of the hexagonal structure. These planes have been identified with the cleavage experimentally using electron diffraction. The weak bonding across these planes is partly due to the greater bond lengths between the lanthanide and coplanar halide ions, relative to the six above- and below-plane ions. Therefore the marked polygons slip vertically intact, with each lanthanide carrying its six nearest neighbors.

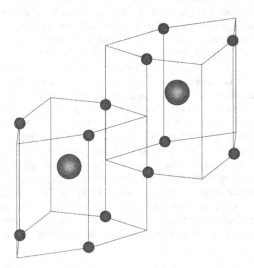

Figure 1. Polygons representing a portion of the LnX_3 structure (c-axis vertical). Both easy slip and cleavage are expected along the shared faces, which represent (11.0) planes in the hexagonal structure

This proposed mechanism is supported by consideration of the elastic strain energy associated with possible Burger's vectors for dislocations. The low c/a ratio indicates roughly 1/3 the energy for c-axis slip, relative to the basal plane. The high packing density and observed low bonding strength of the prismatic planes indicates that {11.0}<00.1> is the primary slip system. Since this Burger's vector is unique in the structure, plastic strain is predicted to be highly anisotropic. Therefore highly localized slip on the prismatic planes would be expected to cause dislocation pileups, building to critical flaws on these planes under relatively low stress. This problem would be exacerbated by the anisotropic expansion, causing shears to develop on these planes for temperature gradients not parallel to the c-axis.

The net effect of this anisotropy is to initiate cleavage fracture in single crystals with extremely low fracture work, as determined quantitatively by fracture toughness estimates. This parameter has been evaluated from microhardness indentations, and plotted against the estimated yield strength in Figure 2 below.

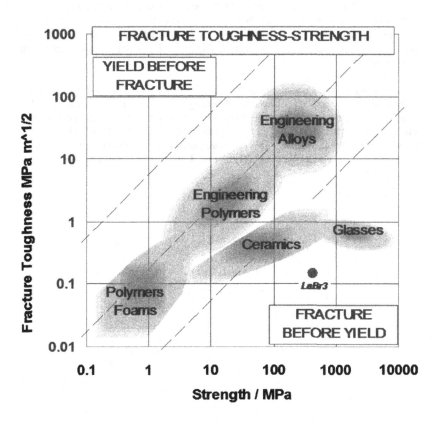

Figure 2. Fracture toughness versus strength for engineering materials and lanthanum bromide. The low toughness of LaBr3 results in cleavage fracture before significant yielding under process stress.

The position of lanthanum bromide on this plot shows that single crystals are extremely fragile, and lie well outside the ranges for useful engineering materials. By contrast, NaI and CsI materials are very ductile and can be extruded or pressed into arbitrary shapes without fracture to form large transparent scintillators. The relative ease of processing these materials is directly related to their cubic crystal structure, giving rise to isotropic thermal expansion and large numbers of independent slip systems.

Strengthening approach

The proposed failure mechanism is initiated by slip on prismatic planes in the LnX3 structure, therefore the fracture toughness in this system is expected to increase in proportion to the yield strength. Methods to strengthen ionic crystals are well known, and their applicability to this system was recently reviewed[6]. Constraints in this application must include minimal alteration of the scintillation process and transparency. This implies that strengthening agents should be incorporated as solid solutions in minimal concentrations. Additives which precipitate or form phases with less than 9-fold coordination of the lanthanide ion will form particles which scatter light, making the crystal cloudy. Therefore the coordination chemistry of potential strengthening agents was considered.

Figure 3 is a plot of coordination number (C.N.) versus anion/cation radius ratio for known AX3 compounds (X = halogen). A variety of crystal structures are observed in these compounds, with C.N. ranging from 3 to at least 11. The plot indicates potential strengthening agents should have a radius ratio below 2 to maintain the 9-fold coordination of the lanthanide. This constraint actually limits the magnitude of the strengthening, since one effect is purely due to elastic distortions produced by the substituents.

4

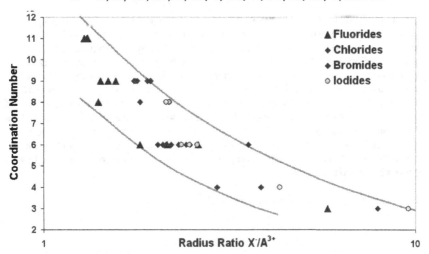

Figure 3. Observed coordination of AX_3 halides showing the trend of decreasing C.N. with increasing anion/cation radius ratio. Cations are listed in order of increasing radius.

The magnitude of the strengthening effects for three known approaches is documented in the literature for a number of ionic crystals. The simplest approach is substitution of one of the matrix ions with a different ion of identical charge; Isovalent substitution. Isovalent alloying introduces elastic distortions in the lattice due to differences in radii, causing drag forces on dislocation. As noted above, this effect also has the potential to cause phase changes resulting in precipitation. Systems which avoid this problem will require larger concentrations of the strengthening agent, since isovalent substitution relies principally on elastic distortions. The concentration dependence of the yield strength in such systems is typically parabolic, with the maximum strength near 50% by mole.

However, a much more powerful strengthening effect is known for substituent ions of a different charge; Aliovalent substitution. Effects caused by aliovalent substitution include introduction of vacancy concentrations far in excess of equilibrium (due to the requirement of charge neutrality), formation of point defects complexes, tetragonal lattice distortion, and coulombic interaction with charged dislocation cores, in addition to elastic distortions. As should be expected, the energetic price of such substituents results in low solubility limits, which often prevents incorporating large concentrations. However, the magnitude of the strengthening effect is quite large, even for small concentrations; typical aliovalent systems exhibit a square root dependence of yield strength on solute concentration, and order of magnitude increase with parts-per-thousand concentrations are common.

Literature results for the above effects in ionic crystals are shown in Table I. The systems employing isovalent substitution show parabolic strengthening, with critical resolved shear stress τ peaking at over 10 near the 50% alloy composition. Systems having miscibility gaps were also investigated, and the maximum increase in yield strength is due to particle-dislocation interactions in this case. These authors reported cloudy crystals for compositions in the gap. By contrast, the aliovalent systems show factors of 6 to 18 increases in room-temperature strength for concentrations as low as 100 ppm. An interesting case is the Al_2O_3-Ti system, in which Pletka studied both Ti(III) and Ti(IV) additions of equal mole %. The isovalent Ti(III) ion resulted in no significant increase in the yield point, whereas the aliovalent Ti(IV) increased the strength greater than a factor of 2. Isovalent Cr(III) ion produced only a modest increase in strength with 4 times higher concentration.

Table I. Strengthening methods in ionic crystals

System	Mechanism	τ/τ_0	Conc.	Dependence	Ref
KCl-KBr	Isovalent substitution	> 10	0-100%	C(1-C)	Katoaka
NaBr-KBr	Isovalent substitution	> 10	0-100%	C(1-C)	Katoaka
NaCl-AgCl	Precipitation	> 10	15-75 %	Misc. gap	Stokes
NaCl-KCl	Precipitation	~ 2	0.5 – 12 %	Misc. gap	Wolfson
NaCl- Ca(II)	Aliovalent subst.	12 - 18	500-900 ppm	$C^{1/2}$	Chin
KCl- Ba(II)	Aliovalent subst.	6 - 15	100-1000 ppm	$C^{1/2}$	Chin
NaBr- Sr(II)	Aliovalent subst.	9 - 18	900-1600 ppm	$C^{1/2}$	Chin
Al2O3-Ti(IV)	Aliovalent subst.	2 (1520 °C)	470 ppm	$C^{1/2}$	Pletka

T. Katoaka and T. Yamada, Japn. J.A.P. 16, 1119 (1977)
R. J. Stokes and C. H. Li, Acta Met., 10, 535 (1962)
G. Y. Chin, et al., J. Am. Ceram. Soc., 56 (1973)
J. B. Pletka, et al., Acta Met., 30,147 IO (1982)
R. G. Wolfson, et al., J. Appl. Phys. 37,704 (1966)
A. Dominguez-Rodriguez et al., J. Am Ceram Soc, 69,281 (1986)

CONCLUSIONS
Cerium and Lanthanum halides form isomorphous alloys with the UCl3 prototype structure. Crystals are highly anisotropic with respect to thermal expansion, plasticity and optical properties, and show perfect cleavage on prismatic planes. Mechanical measurements indicate low fracture toughness, modulus, and yield strength of LaBr3 single crystals, and our analysis of the failure mechanics indicates easy slip in the c direction redistributes residual process stress on prismatic planes, causing cleavage. Cleavage planes are densely packed, and contain the energetically favored Berger's vector, indicating slip and cleavage planes are identical. The limited ductility and

profusion of cracking in this system can therefore be attributed to peculiarities of the crystal structure. Work to apply solid solution strengthening methods in collaboration with Kansas State University is in progress, and further work on mechanical properties to determine elasticity constants, slip systems and high temperature deformation mechanisms to enable predictive modeling of stress and fracture during processing is in progress, in collaboration with Washington State University. Sandia will also continue to research thermodynamic and thermophysical properties, determine phase diagrams, solubility limits, and segregation of strengthening agents.

ACKNOWLEDGEMENT
Project funded by the U.S. Department of Energy NA22, National Nuclear Security Administration. Sandia is a multiprogram laboratory operated by Sandia Corporation, a Lockheed Martin Company, for the United States Department of Energy's National Nuclear Security Administration under contract DE-AC04-94AL85000.

REFERENCES

[1] O. Guillot-Noel et al. / Journal of Luminescence 85 (1999) 21,35

[2] C.W.E. van Eijk / Nuclear Instruments and Methods in Physics Research A 471 (2001) 244–248

[3] BrilLanCe380® from Saint-Gobain Crystals and Detectors, http://www.detectors.saint-gobain.com

[4] Structure and properties of lanthanide halides, F. P. Doty, Douglas McGregor, Mark Harrison, Kip Findley, Raulf Polichar, Proc. SPIE 670705 (2007)

[5] Fracture and deformation behavior of common and novel scintillating single crystals K. O. Findley, J. Johnson, D. F. Bahr, F. P. Doty, and J. Frey, Proc. SPIE 6707, 670706 (2007)

[6] Initial investigation of strengthening agents for lanthanide halide scintillators M. J. Harrison and F. P. Doty, Proc. SPIE 6707, 67070B (2007)

Mater. Res. Soc. Symp. Proc. Vol. 1038 © 2008 Materials Research Society   1038-O02-01

Applications of First Principles Theory to Inorganic Radiation Detection Materials

David Joseph Singh, H. Takenaka, G. E. Jellison, Jr., and Lynn A. Boatner
Materials Science and Technology Division and Center for Radiation Detection Materials and Systems, Oak Ridge National Laboratory, 1 Bethel Valley Rd, Oak Ridge, TN, 37831-6114

ABSTRACT

Applications of first principles methods to understand properties of several known and potential scintillators for radiation detection are described. These include results for rare earth and Pb-based phosphates, rare-earth trihalides, ZnO, perovskites and tungstates.

INTRODUCTION

The development of effective systems for radiation detection applications is an ongoing challenge that is constrained by materials performance. Much research in radiation detection materials has focused on issues of crystal growth and perfection since these are clearly crucial. However, much can also be gained from understanding of trends within different materials families and the specific limitations on performance. These issues can be addressed in part using first principles calculations. Such calculations provide a microscopic window into materials properties that can give chemical insight. This can be of use in finding better materials and guiding modifications of existing materials [1-6]. Here we illustrate this by presenting recent results obtained by applying first principles theory to scintillators used in radiation detection.

The calculations reported here were performed using the general potential linearized augmented planewave (LAPW) method [7], mainly within the local density approximation (LDA). This method provides a highly accurate solution of the density functional equations even for open crystal structures with heavy elements. One of the key inputs for such calculations is the crystal structure. However, in complex scintillators there is often uncertainty about the crystal structure, especially for the internal coordinates. This typically arises in scintillators containing both heavy and light elements, which represents a difficult case for x-ray refinement. In such cases, it is possible to obtain the internal coordinates by minimization of the density functional total energy [8,9]. This was done when needed in the examples discussed below.

ELECTRONIC STRUCTURES

One of the most useful ways of understanding trends in properties of materials is via the electronic structure. Moreover, in scintillators, electronic structure is directly related to scintillator performance, particularly activation and energy transport. We begin with a brief overview of work on the electronic structure of Pb and non-Pb containing phosphates [4].

Phosphate Scintillators: Activation of Pb-Based Glasses

Various phosphates such as $LuPO_4$ can be readily activated with Ce^{3+} and other rare earths to produce good crystalline scintillators. While there are related Pb-based orthophosphate and pyrophosphate materials, it is not known whether these can be the basis of good scintillators or how this might be achieved.

This question is of interest because the Pb-based materials are dense and exceptionally stable from a chemical point of view. However, most importantly the pyrophosphates can be made into highly perfect glasses over wide composition ranges [10,11]. Thus, if these Pb-based phosphate glass materials can be activated, they might form the basis of a family of useful glass scintillators. Motivated by this, we did electronic structure calculations in order to understand the relationships between the Pb and non-Pb containing phosphates. The calculations were done for $ScPO_4$, YPO_4, and $LuPO_4$ as well as Pb containing phosphates. The calculated band structure of $LuPO_4$ [4] is shown in Fig. 1.

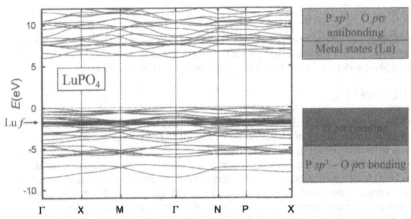

Figure 1. Band structure and schematic depiction of the electronic structure of $LuPO_4$ (Ref. [4]).

The valence bands consist of bonding states of the phosphate groups. The corresponding antibonding states are high in the conduction bands, as shown schematically on the right side of the figure. The essential point is that the fundamental gap is of charge transfer character, since the states at the conduction band edge are of mainly Lu character. The calculated direct band gap is 6.1 eV, which as usual in LDA calculations, is expected to be an underestimate. YPO_4 shows a similar electronic structure with an LDA gap of 5.8 eV, while experimentally cathode-luminescence and scintillation have been observed in $YPO_4:Nd^{3+}$ up to 6.7 eV [12,13], implying a gap of at least 6.7 eV.

The calculated electronic structure of triclinic $Pb_2P_2O_7$ as represented by the projected density of states is shown in the left panel of Fig. 2. As may be seen it is qualitatively similar to the rare earth orthophosphates in that the valence bands are formed from bonding states associated with the phosphate groups, with the corresponding antibonding states high in the conduction bands and a charge transfer gap. In the case of the Pb compounds the gap is to a Pb $6p$ derived manifold of conduction bands between ~ 4 and 7 eV. The presence of these low lying Pb conduction states is the key difference between the Pb and non-Pb based phosphates. The lower lying position of the Pb states relative to the d states of Y, Sc and Lu is related to the ionic properties (s-d and s-p splittings and electronegativity) [4]. Therefore the presence of these low lying Pb $6p$ conduction bands, found in the ortho- and pyrophosphate, is expected to be generic to the chemistry of Pb based phosphates. As mentioned, LDA band gaps are usually underestimates, and so validation is important. This was done using optical absorption measurements for YPO$_4$ and both glass and crystalline Pb$_2$P$_2$O$_7$. The results are shown in the right panel of Fig. 2. The implication of these results is that if the Pb-based materials are to be made into scintillators, methods for activation different from those used in other phosphates will need to be found. This is because the band gaps will invariably be too small for effective activation by Ce^{3+} or other trivalent rare earths.

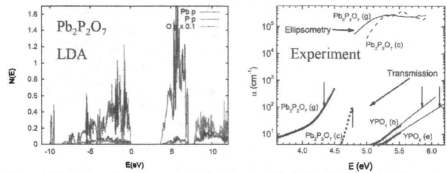

Figure 2. Projections of the electronic density of states of Pb$_2$P$_2$O$_7$ (left) and optical absorption measurements for glass and crystalline Pb$_2$P$_2$O$_7$ and YPO$_4$ following Ref. [4]. The labels (e) and (o) are for polarization parallel and perpendicular to the tetragonal c-axis of YPO$_4$, while (c) and (g) refer to crystal and glass. "LDA" denotes our density functional calculations, while "Experiment" denotes optical absorption measurements.

One interesting feature of the phosphates is that both the valence and conduction bands are broad and dispersive near the band edges (note that because of the number of atoms in the cell the Brillouin zones are small). While it is common for conduction bands to be broad since they arise from spatially extended unoccupied orbitals, this is not so usual for valence bands, and in fact in many semiconductors and insulators holes are heavier and have lower mobility than electrons. The dispersion of the bands in the phosphates is due to the fact that the phosphate anions, whose bonding states form the valence bands, are quite large compared to the rare earth and Pb cations. This is also the case in many good scintillators, such as rare earth trihalides, tungstates and NaI.

Electronic Structure and Energy Transport in LaBr$_3$

The crystal structure [14] of hexagonal LaBr$_3$ is remarkable in this regard. This material, when activated with Ce^{3+} is one of the best known scintillators [15]. The nearest neighbor La-Br distance is 3.09 Å, while the nearest Br-Br distance is 3.58 Å. For comparison, the sum of La^{3+} and Br^{1-} ionic radii is 3.1 Å and twice the Br^{1-} ionic radius is 3.64 Å [16]. Thus in the LaBr$_3$ structure the Br anions are squeezed together, which may be expected to favor hole mobility. As may be seen from the calculated LDA band structure (Fig. 3), this is indeed the case. The valence bands have Br p character as expected, while the conduction bands have metal character. The LDA band structure also shows a narrow set of bands in the gap. This is the La f resonance, which is misplaced here in the LDA. Despite the large band gap, there are strong dispersions in the valence bands near the band edge, though these dispersions are still smaller than the conduction band dispersions. Also, it may be noted that the dispersion along the c-axis direction (Γ-A) is stronger than along the in plane direction. This implies better energy transport along the c-axis.

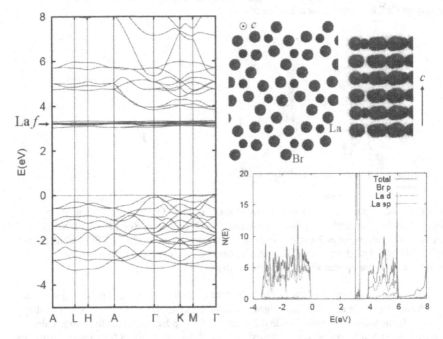

Figure 3. Calculated LDA band structure (left) and projections of the density of states (bottom right) of LaBr₃. These results were obtained using the experimental crystal structure (top right).

Electronic Structure of Li₂WO₄ and Relationship with PbWO₄

The tungstates, e.g. CdWO₄ and PbWO₄ [17,18] form another useful class of scintillators. These are nominally based on $(WO_4)^{2-}$, though it may be more realistic to consider these materials as made up of O^{2-} anions with W^{6+} and other cations, especially considering that both scheelite and wolframite type phases scintillate, while the W is tetrahedrally coordinated in the former and octahedral coordinated in the later. The related phase Li₂WO₄ has been reported [19] and is based on WO₄ tetrahedra, similar to PbWO₄ (it is not, however, in the scheelite structure). This may be of interest for neutron detection because of its high Li content, but it is not known whether this phase is a scintillator. Accordingly, we did electronic structure calculations to determine its relationship with the tungstate scintillators, which have been studied previously [2].

The crystal structure of Li₂WO₄ is based on phenacite and is quite complex with six formula units (42 atoms) per rhombohedral cell and four symmetry distinct O sites and two distinct Li sites. This poses a very challenging problem for structure refinement. So not surprisingly we find that the light atom coordinates in the reported structure are not accurate.

This is in the sense that we find large forces on the atoms in the LDA with this structure. Accordingly we determined the internal coordinates by total energy minimization. These are as given in Table I.

This crystal structure has both Li and W in tetrahedral coordination by O. The electronic structures of $PbWO_4$ and Li_2WO_4 as obtained within the LDA are compared in Fig. 4. In both compounds the valence bands are of O p character, while the conduction bands are mainly of W d character, with additional Pb p character in the case of $PbWO_4$. Also, in both compounds there is substantial hybridization between O p states and W d states, as might be expected from the high valence state of the W^{6+} ions. This is evident in the W d character at the bottom of the O p bands. This arises from states of $t_{2g} - p$ σ bonding character. It is also evident in the case of Li_2WO_4 in the large $e_g - t_{2g}$ crystal field splitting of ~2 eV (the t_{2g} manifold in a tetrahedral ligand field are pushed up because they are the σ antibonding $p - d$ combinations, while the e_g states participate in weaker π interactions).

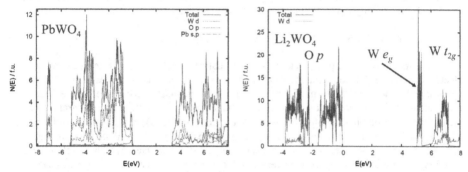

Figure 4. LDA electronic densities of states for $PbWO_4$ (left) and Li_2WO_4 (right).

However, despite the above mentioned similarities in the local bonding, the density of states of Li_2WO_4 is strikingly different from that of $PbWO_4$ and also the other scheelite and wolframite tungstate scintillators. The LDA band gap of Li_2WO_4 is ~2 eV larger than in the other compounds. This is not because the center of the W d states occur at higher energy, but rather because the bands are very much narrower. The same is true of the valence bands. This difference is structural in origin.

Table I. Calculated atom positions of Li_2WO_4. The lattice parameters (R-3: 148), a=8.888Å, α=107.78° are from experiment [19] and the coordinates are given in terms of the rhombohedral setting lattice vectors.

	x	Y	z
W	0.0338	0.4456	0.2714
O1	0.1521	0.4582	0.1421
O2	0.9192	0.5767	0.2494
O3	0.8812	0.2210	0.1846
O4	0.1869	0.5296	0.4961
Li1	0.3701	0.7756	0.6098
Li2	0.7024	0.1055	0.9343

In the crystal structure of PbWO$_4$ [20], the Pb atoms are each coordinated by eight O atoms, from four neighboring WO$_4$ units. This leads to short distances of 2.91 Å between O in different WO$_4$ units to form a three dimensional connected network. This is short enough for direct O-O hopping between the units. This also broadens the W derived conduction bands because of the strong W-O hybridization. In Li$_2$WO$_4$ both the W and Li are tetrahedrally coordinated with O. This leads to less connected WO$_4$ units with a smaller number of near O and longer distances between them. The shortest O-O bond length between WO$_4$ units is 2.99 Å and these bonds do not form a connected network of WO$_4$ units (in the Li$_2$WO$_4$ structure a connected network is formed only when O-O distances of 3.10Å are included).

In any case, the electronic structure of Li$_2$WO$_4$ is found to be very different from the known tungstate scintillators in that the gap is predicted to be larger (note that the experimental gaps will be larger than the LDA gaps in both compounds), and the dispersion of the bands much weaker. The weaker dispersion of the bands in Li$_2$WO$_4$ is expected to be unfavorable for energy transfer. Also the large difference in band gap may mean that the radiative recombination centers in Li$_2$WO$_4$ may be different from those in the other tungstates.

DEFECTS AND ENERGY TRANSPORT

As mentioned, transport is a key issue in radiation detection. This is the case both for electronic radiation detection materials, where charge is collected at electrodes, and scintillators, where energy, mostly in the form of excited charge carriers, must be transferred to the radiative recombination sites where light is produced. In both cases, the relevant materials parameters are the mobility – recombination lifetime (non-radiative for scintillators) products, $\mu_e\tau_e$ and $\mu_h\tau_h$ for electrons and holes, respectively. However, electronic materials and scintillators are in very different regimes. In the electronic materials, carriers must be collected over the whole sample, with length scales of millimeters or more, while for scintillators the relevant length scale is the distance between activator sites, typically ~10-20 Å. In addition, in scintillators, it is important that the crystals be optically perfect with little light scattering or absorption. Understanding the defects that limit performance is important both in electronic materials and in scintillators. First principles calculations can play a useful role in this, both by identifying the types of defects that are important, and also in understanding their nature. Here we illustrate this with some recent results [21] on ZnO and the perovskites, YAlO$_3$ (YAP) and LuAlO$_3$ (LuAP) [22].

Hydrogen in ZnO Defects: Finding of Anionic Character in O Vacancies

ZnO is a much investigated material that is an extremely fast inorganic scintillator when doped with trivalents such as Ga and may be of interest as an oxide electronic material. One important scintillator application is as an α detector in the context of neutron generators [23-26]. For this application, hydrogen treatment has been shown to improve properties. In fact hydrogen treatments are useful for passivating and compensating defects in a variety of traditional non-oxide semiconductors. This depends on the details of the bonding of H in defects.

In this context, since O vacancies may be important in ZnO, density functional calculations were carried out by Janotti and Van de Walle [27] for H in O vacancies. They claimed, based on an analysis of their calculations, that H forms a previously unknown type of

14

multicenter bond with the neighboring Zn atoms in ZnO, and that this bonding type can also occur in other oxides. One advantage of the LAPW method is that it allows a more atom centered analysis of the bonding than planewave type methods, and accordingly we applied such calculations to re-examine H in O vacancies in ZnO [?1].

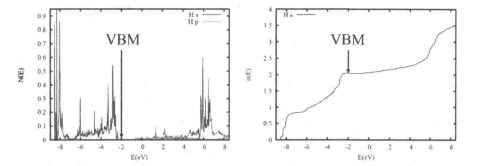

Figure 5. Calculated H projection of the density of states of a neutral 72 atom ZnO supercell with H in an O vacancy (left) and the integrated H contribution to the density of states normalized for the sphere radius (right) following Ref. [21]. The notation "VBM" denotes the energy of the ZnO valence band maximum.

As may be seen in the electronic structure (Fig. 5), hydrogen contributes an s level near the bottom of the valence bands. The H s character is hybridized with other valence band states. However, little admixture with conduction bands is found, and the integral over the occupied valence bands corresponds to very nearly 2 e per H. There is an additional H derived state at ~8 eV above the valence band maximum, but this is associated with the H resonance. Noting that hybridization between occupied states does not constitute bonding, we conclude that there is no prominent multicenter bond involving H coordinated by Zn, and that in fact H occurs as a very stable H⁻ anion and as such compensates half the charge of the O vacancy [21].

Antisite Defects in YAlO₃ and LuAlO₃: Trap States Associated with Al$_Y$ and Al$_{Lu}$

The perovskites YAlO₃ and LuAlO₃ activated with Ce^{3+} (YAP and LuAP, respectively) are remarkable scintillators [28-32]. This is because YAP is reported to be an extremely proportional scintillator [31,32]. This is important because non-proportionality and light yield are the two key materials factors limiting the energy resolution of scintillators. If the light yield could be improved, YAP might become one of the best materials for spectroscopic γ detection. Furthermore, studying YAP may yield insights into how non-proportionality can be reduced in scintillators.

Atomistic simulations [33-35] based on empirical potentials found that a particular anti-site type defect in which an Al and an A-site (Y or Lu) atom are interchanged (denoted here Al$_Y$Y$_{Al}$ or Al$_{Lu}$Lu$_{Al}$) is a low energy defect and may be responsible for observed electron traps in these materials. As such, we did first principles calculations to assess the energetics directly, and to consider other possible anti-site type defects and their possible role as electron traps [22]. The calculations were done directly with relaxed 40-atom perovskite supercells and with energetics

15

computed relative to reservoirs of the binary oxides, Y_2O_3, Al_2O_3 and Lu_2O_3. This is appropriate for oxidizing growth conditions. The calculated energies are as given in Table II.

TABLE II. Calculated LDA energies of anti-site type defects in $YAlO_3$ and $LuAlO_3$.

	$YAlO_3$ (eV)	$LuAlO_3$ (eV)
$Al_{Y(Lu)}$	2.5	2.1
$Y(Lu)_{Al}$	2.4	1.6
$Al_{Y(Lu)}Y(Lu)_{Al}$	4.4	3.4

The energies of anti-site defects consisting of A-site – B-site exchange, considered previously [33-35], are somewhat larger (by ~ 1 eV) than those obtained previously, and are sufficiently large that it seems unlikely that a large enough concentration of these to explain the observed electron traps can occur under the synthesis conditions (note that the Ce^{3+} concentrations in scintillator applications are ~0.1% or larger). However, the single anti-sites are lower in energy and therefore may be more prevalent. The calculated electronic structures show modifications at the conduction band edge for all of these anti-site type defects. However, the largest effects are for defects involving Al on the A-site, in which case clear traps are seen 0.4 – 0.8 eV below the band edge in both YAP and LuAP. This suggests that the properties of LuAP and YAP might be improved by growing under conditions that disfavor Al_{Lu} and Al_Y, e.g. slightly yttria/lutetia rich.

SUMMARY AND CONCLUSIONS

The development and perfection of materials for radiation detection systems remains a challenging materials problem. Much progress continues to be made with improved growth techniques and testing of new scintillator materials. As illustrated by the examples above, first principles calculations are yielding useful insights into the properties of scintillators and their chemical trends. It is expected that the usefulness of this type of approach in conjunction with experimental efforts will grow with the increasing performance of computers and ongoing development of better computational algorithms and codes.

ACKNOWLEDGMENTS

We are grateful for helpful discussions with Z.W. Bell and J.S. Neal. This work was supported by the Department of Energy, Office of Nonproliferation Research and Development (NA-22).

REFERENCES

1. S.E. Derenzo and M.J. Weber, Nucl. Inst. Meth. Phys. Res. A **422**, 111 (1999).
2. Y. Abraham, N.A.W. Holzwarth, and R.T. Williams, Phys. Rev. B **62**, 1733 (2000).
3. M. Klintenberg, S.E. Derenzo, and M.J. Weber, Nucl. Inst. Meth. Phys. Res. A **486**, 298 (2002).
4. D.J. Singh, G.E. Jellison, Jr., and L.A. Boatner, Phys. Rev. B **74**, 155126 (2006).

5. M. Stephan, M. Zachau, M. Groting, O. Karplak, V. Eyert, K.C. Mishra, and P.C. Schmidt, J. Lumin. 114, 255 (2005).
6. V. Lordi, D. Aberg, P. Erhart, and K.J. Wu, Proc. SPIE 6706, 670600-1 (2007).
7. D.J. Singh and L. Nordstrom, Planewaves Pseudopotentials and the LAPW Method, 2nd Edition (Springer, Berlin, 2006).
8. M. Suewattana, D.J. Singh and M. Fornari, Phys. Rev. B 75, 172105 (2007).
9. M.D. Johannes and D.J. Singh, Phys. Rev. B 71, 212101 (2005).
10. B.C. Sales and L.A. Boatner, Mater. Lett. 2, 301 (1984).
11. W.F. Krupke, M.D. Shinn, T.A. Kirchoff, C.B. Finch, and L.A. Boatner, Appl. Phys. Lett. 51, 2186 (1987).
12. V.N. Makhov, N.Y. Kirikova, M. Kirm, J.C. Krupa, P. Liblic, A. Lushchik, C. Luschik, E. Negodin, and G. Zimmerer, Nucl. Inst. Meth. Phys. Res. A 486, 437 (2002).
13. D. Wisniewski, S. Tavernier, P. Dorenbos, M. Wisnewska, A.J. Wojtowicz, P. Bruyndonckx, E. van Loef, C.W.E. van Eijk, and L.A. Boatner, IEEE Trans. Nucl. Sci. 49, 937 (2002).
14. K. Kraemer, T. Schleid, M. Schulze, W. Urland, and G. Meyer, Z. Anorg. Chem 575, 61 (1989).
15. E.V.D. van Loef, P. Dorenbos, C.W.E. van Eijk, K. Kramer, and H.U. Gudel, Appl. Phys. Lett. 79, 1573 (2001).
16. R.D. Shannon, Acta Cryst. A32, 751 (1976).
17. P. Lecoq et al., Nucl. Inst. Meth. Phys. Res. A 365, 291 (1995).
18. P. Lecoq, Nucl. Inst. Meth. Phys. Res. A 537, 15 (2005).
19. W.H. Zachariasen and H.A. Plettinger, Acta Cryst. 14, 229 (1961).
20. R. Chipaux, G. Andre, and A. Cousson, J. Alloys Comp. 325, 91 (2001).
21. H. Takenaka, and D.J. Singh, Phys. Rev. B 75, 241102 (2007).
22. D.J. Singh, Phys. Rev. B 76, 214115 (2007).
23. W. Lehmann, Solid-State Electron. 9, 1107 (1966).
24. D. Lucky, Nucl. Inst. Meth. 62, 119 (1968).
25. J.C. Cooper, D.S. Koltick, J.T. Mihalczo, and J.S. Neal, Nucl. Inst. Meth. Phys. Res. A 505, 498 (2003).
26. J.S. Neal, L.A. Boatner, N.C. Giles, L.E. Haliburton, S.E. Derenzo, and E.D. Bourret-Courchesne, Nucl. Inst. Meth. Phys. Res. A 568, 803 (2006).
27. A. Janotti and C.G. Van de Walle, Nat. Matter. 6, 44 (2007).
28. M.J. Weber, J. Appl. Phys. 44, 3205 (1973).
29. V.G. Baryshevsky, M.V. Korzhik, B.I. Minkov, S.A. Smimova, A.A. Fyodorov, P. Dorenbos, and C.W.E. van Eijk, J. Phys. Condens. Matter 5, 7893 (1993).
30. A. Lempicki, M.H. Randles, D. Wisniewski, M. Balcerzyk, C. Brecher, and A.J. Wojtowicz, IEEE Trans. Nucl. Sci. NS-42, 280 (1995).
31. M. Kapusta, M. Balcerzyk, M. Moszynski, and J. Pawelke, Nucl. Inst. Meth. A 421, 610 (1999).
32. W. Mengesha, T.D. Taulbee, B.D. Rooney, and J.D. Valentine, IEEE Trans. Nucl. Sci. NS-45, 456 (1998).
33. M.M. Kukja, J. Phys. Condens. Matter 12, 2953 (2000).
34. C.R. Stanek, M.R. Levy, K.J. McClellan, B.P. Uberuaga, and R.W. Grimes, Phys. Stat. Sol. (b) 242, R113 (2005).
35. C.R. Stanek, K.J. McClellan, M.R. Levy, and R.W. Grimes, J. Appl. Phys. 99, 113518 (2006).

Mater. Res. Soc. Symp. Proc. Vol. 1038 © 2008 Materials Research Society   1038-O03-01

Spontaneous Increase of Electric Conductivity of Nanostructure Dielectrics

M. P. Lorikyan
Armenia Yerevan Physics Institute, 2 Alikhanian Br. Str., Yerevan, 0036, Armenia

Introduction

It is well known that the surface-to-volume ratio of nanoparticles is very large and a significant fraction of atoms or ions are close to the surface of nanoparticles. On the other hand, secondary electron emission (SEE) occurs at or close to the surface of emitters, and, as a result, SEE is very sensitive to the surface properties of materials. Therefore, in the materials with high surface-to-volume ratio, secondary electrons readily get out from the surface and, hence, nanostructure materials may have high SEE characteristics. Actually, it was found in 1972 that under the influence of an external electric field, a high efficient controllable drift and multiplication (CEDM) of electrons takes place in porous KCl and CsI [1,2,3]. It has also been shown that because of polarization of these materials, the CEDM process becomes noticeably suppressed and unstable [4-9]. However, recent detailed investigations have shown that for porous CsI prepared by thermal evaporation, polarization occurs for a time after deposition, then polarization charges spontaneously vanish over time and porous CsI does not display any signs of polarization [10-12] (i.e., CEDM exhibits a high gain, high stability and good spatial localization). Thus, porous CsI spontaneously looses the ability to polarize with time. It was also shown that the time T_{pl}, during which the spontaneous polarization 'faculty' loss phenomenon (SPFL) takes place, strongly depends on the temperature T_k, at which porous CsI is kept after its preparation. When, after preparation, the porous CsI is kept at $T_k=18^0C$, SPFL phenomenon takes place within a few hours, whereas at $T_k=31^0C$ T_{pl} is at most 1 h [12]. It is very important that after the loss of the polarization ability, porous CsI does no longer exhibits polarization, regardless of temperature, or whether or not it is under voltage and/or ionizing radiation. Thus, immediately after thermal preparation of porous CsI, accumulation of charges starts in it under the action of CEDM, but then these polarization charges spontaneously vanish with time and do not accumulate again, thus ceasing to behave as a dielectric. Later, Chianell et al. [13] observed that the structure of porous CsI used in [1-12] has the form of whiskers 10–50 nm in diameter, i.e. porous CsI has nanostructure.

Note that it has been known for a long time that some dielectrics with porous structure are characterized by high-gain electron multiplication in an electric field of positive charges accumulated on their surface, but this effect is not controllable and inertial (Molter effect) [14-21].

In this article, using identical experimental conditions the phenomenon of SPFL in porous CsBr and CsI is investigated, and the results are compared. Both materials are of the same purity (99.99%). As a method for investigation of the SPFL phenomenon, the CEDM of electron clusters is used.

Experiment and Results

As the cross sectional dimension of electron clusters and stability of CEDM strongly depend on the influence of polarization charges, time stability of CEDM is investigated and the average z is estimated. The clusters are counted in which the number of electrons is higher than the registration threshold n_o. Separately, the numbers of clusters with $z<b = 250\mu m$ ($N_{cl}(s)$), and

$z \geq b = 250$ μm ($N_{cl}(c)$) are counted. In order to evaluate z, an electron cluster detector is used. This detector consists of anodes made of a gold plated wire (25-micron in diameter) stretched over an insulator frame, cathode and porous active material that fills the gap between the anodes and cathode. The distance between the anode wires is b=250μm. An electric field is applied between the anodes and cathode. 5 MeV α-particles enter into the porous medium from the anode wire side, pass through the porous dielectric and create electron clusters in the porous medium. The cluster electrons are accelerated in the pores by the applied electric field. The accelerated electrons produce new secondary electrons in the pore walls. This process takes place for all generations of secondary electrons, and if the average number of secondary electrons emitted after each electron-wall impact is greater than one, an avalanche multiplication of electrons arises. The electron and hole clusters move under the influence of the external electric field in opposite directions, and a negative pulse arises on the anode.

To determine z, the anode wires are divided into two groups. The first group includes the even-numbered and the second group the odd-numbered anode wires. The anode wires in each group are connected with each other, and the pulses from each group are counted separately. Both $N_{cl}(s)$, the number of clusters detected by one wire of the anode wire groups (z<b), and $N_{cl}(c)$, the number of clusters detected simultaneously by two adjacent wires of both anode wire groups ($z \geq b$), are counted. The total number of detected clusters is $N_{cl} = N_{cl}(s) + N_{cl}(c)$. It is clear that when z is much smaller than the distance between the anode wires b, the clusters are registered by a single anode wire (by one group), i.e. $N_{cl}(c) << N_{cl}(s)$, and the parameter $S = N_{cl}(c)/N_{cl}(s) << 1$, and when $z \geq b$, the clusters are registered by two adjacent anode wires (by both groups) simultaneously and $N_{cl}(c) \approx N_{cl}$, $S \approx 1$.

The porous layers were prepared by thermal deposition of CsI and CsBr in a low pressure Ar atmosphere [22]. The relative density of CsI and CsBr was $\rho_{rel} \approx 0.4\%$ and the thickness of the layers just after deposition was 0.8 mm. As the gap between the anode and cathode of the cluster detector was 0.5 mm, after assembling the device, the porous layer was compacted by a factor of 1.5-2. After deposition of the porous material, the cluster detector was quickly installed in vacuum chamber filled with argon. The vacuum chamber was then brought to a vacuum of $\approx 7 \times 10^{-3}$ Torr, and the measurements were started.

Fig. 1a shows the change in the number of clusters with the applied voltage V detected by one anode wire only ($N_{cl}(s)$-white circles) and by two adjacent wires ($N_{cl}(c)$-black circles), for porous CsBr. These data were obtained in 1 h after deposition of porous CsBr at $T_k = 24$ ^{0}C. The experiment lasted 18 min. It is seen that $N_{cl}(s)$ and $N_{cl}(c)$ as well as $S=N_{cl}(c)/N_{cl}(s)$ rise sharply with voltage: at V= 850 V $S = 3 \times 10^{-2}$, and at V = 945 V S = 0.5. The reason is clear, the SEE gain and the energy of cluster electrons increase with the electric field strength [2,16], hence both the number of clusters with more than n_0 electrons and the share of the electrons reaching the adjacent wires simultaneously increase. Fig. 1b shows time stability of $N_{cl}(s)$ and $N_{cl}(c)$ ($N_{cl}(s)$-white circles, $N_{cl}(c)$-black circles) immediately after measuring the V dependence (Fig. 1a).

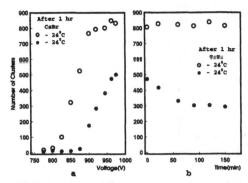

Fig.1 [a] Dependences of $N_{cl}(s)$ and $N_{cl}(c)$ on the applied voltage V for porous CsBr. Data are obtained in $T_{pl} = 1$ h after deposition at $T_k = 24\ ^0C$: white circles-$N_{cl}(s)$, black circles-$N_{cl}(c)$; [b] Time stability of the number of registered clusters for porous CsBr: white circles- $N_{cl}(s)$, black circles-$N_{cl}(c)$.

It follows from Fig.1b that $N_{cl}(s)$ is constant over 160 minutes, whereas $N_{cl}(c)$ decreases with time, i.e. CEDM is stable and the cross sectional dimension of the clusters, z, decreases with time. Measurements carried out the next day (at night the detector was switched off) demonstrate that $N_{cl}(s)$ stability is maintained at the same level as in the beginning, but $N_{cl}(c)$ continues to decrease, thus the improvement in spatial localization of CEDM is also continued when the voltage is switched off. This means that the improvement in the CEDM spatial localization also take place without applied voltage, i.e. spontaneously. The V dependence and time stability measurements carried out repeatedly on a daily basis over a period of 11 days with 18 h interruptions have shown that the CEDM process is stable, but its spatial localization only becomes stable after 2 days.

To more accurately compare the SPFL properties of porous CsBr under the same conditions ($T_{pl} = 1$ h, $T_k = 24\ ^0C$), CEDM was also studied in porous CsI. The V dependences of $N_{cl}(s)$ and $N_{cl}(c)$ are presented in Fig. 2a. White squares correspond to $N_{cl}(s)$, black squares to $N_{cl}(c)$. Fig. 2b shows the time stability of $N_{cl}(s)$ and $N_{cl}(c)$ immediately after obtaining the results presented in Fig. 2a. White squares correspond to $N_{cl}(s)$, black squares to $N_{cl}(c)$. Comparing Fig. 1a with 2a and Fig. 1b with 2b reveals that $S = N_{cl}(c)/N_{cl}(s)$ for porous CsI is always less than that for porous CsBr, i.e., in all cases the spatial localization of electron clusters in porous CsI is higher than in porous CsBr. Thus, within an hour after preparation and later, the density of polarization charges in porous CsI is substantially less than that in porous CsBr. The difference between CsI and CsBr is probably due to the fact that the nanostructure formation process for porous CsBr is longer than that for porous CsI. This effect may also be related to the fact that the CsBr used was considerably older than CsI.

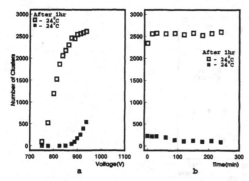

Fig. 2. [a] Dependences of $N_{cl}(s)$ and $N_{cl}(c)$ on the applied voltage V for porous CsI. Data are obtained in T_{pl} =1 h after deposition at T_k = 24 ^{0}C: white squares-$N_{cl}(s)$, black squares-$N_{cl}(c)$; [b] Time stability of the number of registered clusters for porous CsI.: white squares-$N_{cl}(s)$, black squares-$N_{cl}(c)$.

Measurements were also made of CEDM in porous CsI maintained for 1 h and 18 h after preparation at T_k = 15^{0}C (between the measurements the cluster detector is switched off). Fig. 3a presents $N_{cl}(s)$ and $N_{cl}(c)$ at various anode voltages V. The number of clusters registered by one anode wire ($N_{cl}(s)$) only in 1 h after preparation is presented by white circles, and the number of clusters registered by two adjacent wires ($N_{cl}(c)$) by black circles. The number of clusters registered by one anode wire ($N_{cl}(s)$) only in 18 h after preparation is presented by white stars, while the number of clusters registered by two adjacent wires ($N_{cl}(c)$) with black stars. As seen from Fig. 3a, 1) 1 h after preparation of porous CsI, $N_{cl}(s)$ (white circles) CEDM is not spatially localized ($N_{cl}(s) \ll N_{cl}(c)$); in the very beginning (just after switching on the detector) $N_{cl}(s)$ grows a little, but then decreases with voltage, while $N_{cl}(c)$ always rises sharply with voltage. 2) 18 h after preparation, the CEDM characteristics are changed dramatically: $N_{cl}(s)$ becomes high and exceeds $N_{cl}(c)$, i.e. CEDM is already spatially localized. It is seen from Fig. 3b that in 18 h after preparation of porous CsI, $N_{cl}(s)$ is very high ($N_{cl}(s) \gg N_{cl}(c)$) and almost stable, $N_{cl}(c)$ is also stable but much lower.

The fact that the number of registered clusters ($N_{cl}(s)$) with cross sectional dimension z<b in the very beginning does not increase with V, indicates that the polarization charges are accumulated in porous CsI. The increase in the number of clusters registered by two adjacent wires $N_{cl}(c)$ is due to accumulation of these charges, which scatter the cluster electrons and increase z.

So, it can be concluded that the nanostructure formation time in porous CsI decreases with increasing temperature at which the porous CsI is kept after thermal deposition.

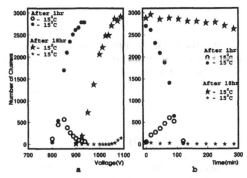

Figure 3 [a] Dependences of $N_{cl}(s)$ and $N_{cl}(c)$ on the applied voltage V for porous CsI. Data are obtained at: 1) $T_{pl} = 1$ h after deposition at $T_k = 15\ ^0C$: white circles-$N_{cl}(s)$, black circles-$N_{cl}(c)$; 2) $T_{pl} = 18$ h after deposition at $T_k = 15\ ^0C$: white stars-$N_{cl}(s)$, black stars-$N_{cl}(c)$. [b] Time stability of the number of registered clusters for porous CsI. Data are obtained at: 1) $T_{pl} = 1$ h after deposition at $T_k = 15\ ^0C$: white triangles-$N_{cl}(s)$, black triangles-$N_{cl}(c)$; 2) $T_{pl} = 18$ h after deposition at $T_k = 15\ ^0C$: white stars-$N_{cl}(s)$, black stars-$N_{cl}(c)$.

Discussion

It can be concluded that in porous CsI and CsBr prepared by thermal deposition, polarization charges accumulate as a result of CEDM, but then after maintaining CsI and CsBr in an inert gas atmosphere, these charges completely vanish with time and thereafter do not accumulate again (under the effect of CEDM), so porous CsI looses its polarization faculty in some degree. The increase of polarization faculty loss time with the temperature at which porous CsI is kept after thermal deposition is not yet understood; possibly, porous CsI becomes nanostructured in a definite time after preparation, and this time depends on temperature.

Discovery of SPFL phenomenon opens new perspectives of the development of novel technologies in the area of SEE emitters, photocathodes, x-ray converters and detectors and different particle detectors, however more detailed investigations are needed. At present it is known that carbon nanotubes have high auto electron emission coefficient [24,25], and nanostructured porous CsI has high secondary electron emission coefficient [10-12]; this work shows that CsBr is also characterized by extremely high secondary electron emission properties. Note that nanostructure porous CsI detectors with high spatial (20 μm) resolution [10-12,30], very fast response time (about 1 ns) and time resolutions of 60 ps [31] have already been developed. As for the physical nature of SPFL phenomenon, high CEDM efficiency and high electron multiplication factors are obviously related to the nanostructure of the materials used. It is known that many electric effects for nanosize objects are related to the fact that when the object size approaches the free-path length, de Broglie wavelength or other parameters characterizing electrons in electron gas, the properties of these objects become dependent on their sizes, and quantum effects appear. These quantum effects change the generation and transfer mechanisms of the charges [26,27,28]. Examples are tunneling and the decrease of the carbon nanotube forbidden zone [29].

23

Acknowledgments

The author expresses his gratitude to G. Ayvazyan and G. Asryan for their assistance in experiments, Professors F. Sauli, A. K. Odian and G. Charles for significant support. The work was supported by the International Science and Technology Centre.

References

[1] Lorikyan M. P. et al., Izv. Akad.Nauk ArmSSR Fiz 6 297 (1967)
[2] Lorikyan M. P. et al., Nucl. Instr. and Meth. **122** (1974) 377
[3] Lorikyan M. P. et al., Nucl. Instr. and Meth. **140** (1977) 505
[4] Gukasyan S. M. et al., Nucl. Instrum. Methods **167** (1979) 427
[5] Gukasyan S. M. et al., Nucl. Instrum. Methods **171** (1980) 469
[6] Lorikyan M. P. et al., Nucl. Instrum. Methods A **340** (1994) 625
[7] Lorikyan M. P. et al., Nucl. Instrum. Methods A **350** (1994) 244
[8] Lorikyan M.P. 1995 Phys.—Usp. **38** 1995 1271–81
[9] Lorikyan M. P. Nucl. Instrum. Methods A **454** (2000) 257
[10] Lorikyan M. P. Nucl. Instrum. Methods A **510** (2003) 150–7
[11] Lorikyan M. P. Nucl. Instrum. Methods A **513** (2003) 394
[12] Lorikyan M. P. Nucl. Instrum. Methods A **515** (2003) 701
[13] Chianell C. et al., Nucl. Instrum. Methods A **373** (1988) 245
[14] Malter L.1936 Phys. Rev. **50** 48
[15] Zernov D. V. 1937 Zh. Eksp. Teor. Fiz. **17** 1787
[16] Stenglass E. J. and Goetze G. W. 1962 IRE Trans. Nucl. Sci. **8** 83
     Stenglass E. J. and Goetze G W. 1962 IRE Trans. Nucl. Sci. **9** 97
[17] Jacobs H. 1951 Phys. Rev. **84** 877
[18] Jacobs H. et al., 1952 Phys. Rev. **88** 492
[19] Garvin E. L. and Edjecumbe J 1969 Preprint SLAC-PUB 156
[20] Llacer J. and Garvin E. L. 1969 J. Appl. Phys. **40** 101
[21] Garvin E. L. and Llacer J. 1970 J. Appl. Phys. **41** 1489
[22] Gavalyan V. G. et al., Izv.Akad. Nauk Arm. SSR, Ser. Fiz. 17 (1982) 102.
[23] Sternglass E. J. 1957 Phys. Rev. **108** 271
[24] Eletski A. B. 1999 Usp. Fiz. Nauk. **167** 945 (in Russian)
[25] Eletski A. B. 2002 Usp. Fiz. Nauk. **172** 401 (in Russian)
[26] Hofmeister H. 1993 Proc. NATO Advanced Study Institute on Nanophase
     Materials: Synthesis–Properties– Applications (Corfu, June–July) p 209
[27] Gulyaev Yu V et al., 1994 7th Int. Vacuum Microelectronics Conf.
     France, July (994) p 322 (Suppl. 271)
[28] De Heer W. A. et al., 1995 Science **270** 1179
[29] Saito R. et al 1998 Physical Properties of Carbon Nanotubes (Singapore:
     World Scientific)
[30] Lorikyan M. P. et al., Nucl. Instrum. Methods A **570 (2007)** 475
[31] **Gavalyan V.G. et al.,** Nucl. Instr. and Meth. A337 (1994) 613.

Mater. Res. Soc. Symp. Proc. Vol. 1038 © 2008 Materials Research Society   1038-O03-02

Organic Semiconductors for Detection of Ionizing Radiation

Tiffany M.S. Wilson[1,2], Douglas A. Chinn[3], Michael J. King[1,4], and F. Patrick Doty[1]

[1]Engineered Materials, Sandia National Laboratories, Livermore, CA, 94550
[2]Chemical and Biomolecular Engineering, The Ohio State University, Columbus, OH, 43210
[3]Sandia National Laboratories, Albuquerque, NM, 87185
[4]Nuclear Engineering, University of California, Berkeley, CA, 94720

ABSTRACT

Organic semiconductors are increasingly common in electronics and sensors, and are now under investigation for a novel type of radiation sensor at Sandia National Laboratories. These materials can offer wide band gaps, high resistivities, low dielectric constants, and high dielectric strengths, suggesting they may be suitable for solid-state particle counting detectors. A range of solution cast materials have been evaluated for this application, primarily in the family of poly(p-phenylene vinylene)s, or PPVs. The high ratio of hydrogen to carbon offers neutron sensitivity, while the low Z material provides some natural gamma discrimination. Compared to existing detectors, these materials could potentially offer large-scale radiation detection at a substantially reduced cost.

While PPVs hold promise for radiation detection, the mechanical and electrical properties must be optimized and the processing effects understood. Polymers can offer significantly simplified processing compared to the more common crystals used in solid state detection, which can be size limited and fragile. However, organic semiconductors are very sensitive to processing conditions, and mobility can be affected by orders of magnitude by processing variables, without altering any covalent chemistry. Additives can also have dramatic effects on both electrical and mechanical properties. We report on nanoparticle additives that cause an increase in photoresponse of approximately three orders of magnitude as compared to a polymer film without additives. We separately show an order of magnitude increase in photoresponse by exposing the polymer/fullerene composite to sub-bandgap light.

Future work will analyze the feasibility of single particle detection and various geometries for optimization. Additional processing variables will also be investigated for further improvement of mobility and reduction of trap density.

INTRODUCTION

Conjugated polymers are being investigated at Sandia National Laboratories for use as detectors of ionizing radiation, in particular fast neutrons (> 0.1 MeV). Conjugated polymers offer room-temperature semiconducting properties for signal detection, but can still be high resistance, providing low leakage current. In addition, many of them have a high dielectric strength and low dielectric constant, allowing them to be used at high bias and with low noise. With careful side chain selection, a high ratio of hydrogen to carbon can be possible, providing good neutron sensitivity. The use of a polymer, as opposed to more conventional crystalline semiconductor materials, opens up many opportunities for decreased cost, increased size, and simplified manufacturing methods. They also offer potential integration with pre-existing systems made possible by conformal coating methods.

A polymer detects an incoming neutron based on the proton recoil reaction, as detailed by Knoll [1]. Since ^{1}H has the maximum fractional energy transfer in a neutron elastic scattering event, this is the ideal detection atom, and thus we want a very high ratio of hydrogen to other elements. Once a proton has been knocked from its original position by the incoming neutron, it creates electron hole pairs as it loses energy. These charges then need to be separated and transported to the electrodes for collection by the applied field. Efficient charge collection relies on the electron and hole mobilities in the material as well as the trap states in the material, which are a strong function of defects. In the case of a polymeric material, these properties are highly dependant on processing. Therefore, this research examines the effects of processing variables and additives on the charge collection properties in π-conjugated polymers. To control for covalent chemistry, the following work was all done with the same polymer, from the family of poly(p-phenylene vinylene)s, PPVs. The specific PPV of focus is poly[2,5-bis(3′,7′-dimethyloctyloxy)-1,4-phenylenevinylene], hereafter referred to as $OC_{10}PPV$, as depicted in Figure 1. This specific polymer was chosen for its high resistivity, high hydrogen content, side chain symmetry, solubility, and commercial availability.

The majority of current photovoltaic research with polymers utilizes the concept of a bulk heterojunction. Rather than providing one uniform heterojunction between materials with different electrical properties, the two are intimately mixed, provided a large interfacial surface area and reduced distance between interfaces, ideally with distances approaching that of the exciton diffusion length. The bulk heterojunction thus works by providing an interface to separate an exciton into mobile charges, which can then be transported through the material to the electrodes. Numerous papers report improved performance and photoconductive response by combination of a hole transporting polymer with an electron withdrawing carbon nanoparticle [2-5]. Lee et al. reported over an order of magnitude increase in steady-state photoconductivity in a 50 wt% composite of a substituted PPV and C_{60} as compared to the pristine polymer [4]. One significant obstacle then becomes the proper ratios of the two components and ideal dispersion or organization of the two components. This in itself is a large research area [6]. Numerous groups report improved solubility and dispersion of fullerene-based nanoparticles by adding a binding group to the fullerene, such as with PCBM (methanofullerene [6,6]-phenyl C_{61}-butyric acid methyl ester), the most commonly used such particle. This particle will therefore be utilized in some of our research and is depicted in Figure 1.

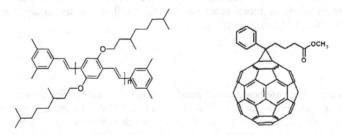

Figure 1. Chemical structure of $OC_{10}PPV$ (left) and PCBM (right)

Amorphous semiconductors are known for having a high density of trap states, thus limiting their electrical performance [7, 8]. This also points to the possibility of exposing the polymer to sub-band-gap light in order to increase transitions from localized band gap states into delocalized band states, a process known as photo de-trapping. This has been reported widely for inorganic semiconductors [9, 10], and the mechanisms of such transitions have been studied in detail by other groups [8]. The bandgap of PPVs is typically around 2.4 eV [11].

EXPERIMENTAL DETAILS

Different solutions were used for the two tests described. For the nanoparticle tests, polymer and PCBMs were obtained from American Dye Source (ADS) in Quebec, Canada and used without additional processing. The polymer had an average molecular weight of 210,000 as measured by the supplier by gel permeation chromatography. Each was dissolved separately in solvent at equal weight percentages, and then the solutions were combined in a 1:1 ratio in a third vial. The straight polymer solution was used as the control. The composite solution and the polymer solution were each drop cast onto interdigitated gold electrodes (IDEs) with 32 μm spacing between electrodes on a glass substrate. For the excitation source, we used an SRS NL100 337 nm nitrogen laser with a rhodamine dye cell tuned to 580 nm, near the peak excitation determined from previous testing. A pulse frequency of 3 Hz was used, and all data was averaged over at least 150 pulses. Measured laser energy after the dye cell and an iris was approximately 10 μJ. A neutral density filter could then be rotated in to control the laser intensity with a neutral density of 0, 1, or 2.5 leading to a final intensity of less than 1% of the unfiltered laser. The laser light was directed through the glass substrate to excite the polymer film, with a controlled bias across the IDE of between 10^4 and 10^5 V/cm. The signal was sent through a preamplifier to a digitizing oscilloscope where it was integrated for 1500 μs.

For the optical pumping tests, unmodified C_{60} was obtained from SES Research in Houston, Texas, and used instead of PCBMs. The same polymer was used, but from Aldrich instead of ADS, and of a different molecular weight than for the PCBM additive tests. Both were used without modification. Because of the low solubility of unmodified fullerenes, the total fullerene content was held to below 10 wt%. We used a Thermo nitrogen laser with a rhodamine dye cell tuned to 590 nm. An infrared light emitting diode with peak wavelength 940 nm (1.32 eV) was directed at the sample, without blocking the laser, and could be easily turned on and off. This test was performed with a planar test structure of a composite film coated onto an indium tin oxide (ITO) on glass substrate then sputtered with a gold contact on top.

RESULTS AND DISCUSSION

In an undoped, amorphous, π-conjugated polymer sample, the charge collection efficiency is too low to accurately detect single particles, so we are investigating a variety of methods to improve charge collection and thus improve detector accuracy and efficiency. We report here on two parameters that have shown dramatic increases in charge collection for laser excited photoconductivity testing. This test serves as a convenient method to examine excitation in conjugated polymers. We have also shown recently that post-polymerization stretching increases the bulk order in the polymer film, as measured by infrared dichroism, and results in an approximate doubling of charge collection [12]. Recent tests have shown that this stretch alignment is also possible with at least 20 wt% of PCBM blended with $OC_{10}PPV$, resulting in a comparable bulk order to a pure polymer stretched similarly. Higher weight percentages of PCBM were also successfully stretch-aligned, but the mechanical properties of the resulting films prevented an accurate measure of the material order.

Nanoparticle Additives

The addition of PCBM nanoparticles is shown to increase the charge collection in a pulsed laser photoconductivity test by approximately three orders of magnitude, as shown in Figure 2. The data is plotted as the square root of the electric field versus the log of charge collected, and appears nearly linear, following Poole-Frenkel behavior, as expected for a PPV [13]. The jump in the signal for the composite occurs at the point where the neutral density filter was changed. The integrated signal was divided by the adjusted laser intensity to partially account for this, but there was still some degree of nonlinearity. This increase in charge collection is higher than most reported in the literature, which are typically closer to one order of magnitude [4]. This increase could be due to the differences in dispersion of the nanoparticle and morphology of the composite, as compared to that produced by other groups. We should note, however, that the composite films did not have the same mechanical properties as the pure polymer film, so optimization of the weight ratios will have to balance electrical improvements with the mechanical properties necessary for a particle detector.

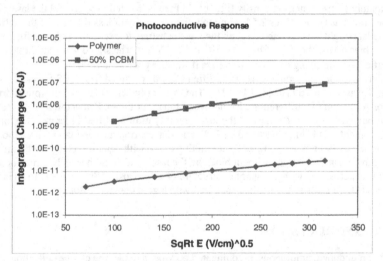

Figure 2. Charge collected across 32 μm interdigitated electrodes in 1500 μs, divided by the laser intensity which was modified with neutral density filters. The lower line (diamonds) represents a film of $OC_{10}PPV$, and the upper line (squares) represents a composite of $OC_{10}PPV$ with 50% PCBM by weight, resulting in an increase of nearly three orders of magnitude.

Optical Pumping

Exposure to sub-band-gap light is shown to increase steady-state photoresponse roughly an order of magnitude, as shown in Figure 3. As labeled in the figure, no response was recorded with the laser shutter closed, and a steady-state response of approximately 1.2 (arbitrary units) was recorded with the laser shutter open, after a settling time of about 10 seconds. Upon exposure to an infrared light source, the signal increased to nearly 20 times its steady state level. The response then decreased over a few minutes in an exponential manner to a steady state response greater than 12. As can be seen in the figure, this exponential trend repeated even when the laser was blocked intermittently. Based on the excitation energy being well below the bandgap, and the slow transient as is common with trap states, this appears to show evidence of photo-detrapping.

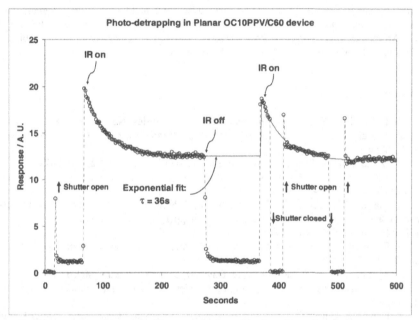

Figure 3. Photoresponse from pulsed laser excitation, showing photo-detrapping upon exposure of an $OC_{10}PPV:C_{60}$ composite film to infrared light. The shutter refers to the laser shutter, not the infrared light.

CONCLUSION

Charge collection efficiency is a major challenge toward the realization of polymer neutron detectors. However, we have shown an order of magnitude improvement with photo-detrapping, and three orders of magnitude improvement by blending with large amounts of modified carbon nanoparticles. By combining and optimizing these two parameters, we may be approaching the necessary values to achieve single particle neutron detection. The addition of other processing methods such as stretch alignment could also aid in the optimization of charge collection. However, this must be carefully balanced with the mechanical properties necessary for a high quality, robust detector system. If successful, this advance could enable large scale, conformable, lightweight, inexpensive neutron detection.

ACKNOWLEDGEMENT

Sandia National Laboratories is a multiprogram laboratory operated by Sandia Corporation, a Lockheed Martin Company, for the United States Department of Energy's National Nuclear Security Administration under contract DE-AC04-94AL8500. This work was supported by Sandia's internal Laboratory Directed Research and Development program and the National Science Foundation under Grant No. 0221678. Additional thanks to Dr. L. James Lee and Dr. Arthur Epstein at The Ohio State University for their guidance and to David Robinson of Sandia National Laboratories for use of his infrared spectrometer.

REFERENCES

[1]   G. F. Knoll, *Radiation detection and measurement*, Wiley, New York **2000**.
[2]   C. J. Brabec, F. Padinger, N. S. Sariciftci, J. C. Hummelen, *Journal of Applied Physics* **1999**, *85*, 6866.
[3]   N. S. Sariciftci, L. Smilowitz, A. J. Heeger, F. Wudl, *Science* **1992**, *258*, 1474.
[4]   C. H. Lee, G. Yu, D. Moses, K. Pakbaz, C. Zhang, N. S. Sariciftci, A. J. Heeger, F. Wudl, *Physical Review B* **1993**, *48*, 15425.
[5]   T. Martens, T. Munters, L. Goris, J. D'Haen, K. Schouteden, M. D'Olieslaeger, L. Lutsen, D. Vanderzande, W. Geens, J. Poortmans, L. De Schepper, J. V. Manca, *Applied Physics a-Materials Science & Processing* **2004**, *79*, 27.
[6]   T. B. Singh, S. Gunes, N. Marjanovic, N. S. Sariciftci, R. Menon, *Journal of Applied Physics* **2005**, 97.
[7]   A. V. Yakimov, V. N. Savvate'ev, D. Davidov, *Synthetic Metals* **2000**, *115*, 51.
[8]   D. Vanmaekelbergh, M. A. Hamstra, L. van Pieterson, *Journal of Physical Chemistry B* **1998**, *102*, 7997.
[9]   S. Kalem, A. Curtis, Q. Hartmann, B. Moser, G. Stillman, *Physica Status Solidi B-Basic Research* **2000**, *221*, 517.
[10]   L. Szaro, *Thin Solid Films* **1985**, *127*, 257.
[11]   P. Posch, R. Fink, M. Thelakkat, H. W. Schmidt, *Acta Polymerica* **1998**, *49*, 487.
[12]   T. M. S. Wilson, F. P. Doty, D. A. Chinn, M. J. King, B. A. Simmons, *SPIE Proceedings*, San Diego, CA, USA, **2007**, 670710.
[13]   S. V. Rakhmanova, E. M. Conwell, *Synthetic Metals* **2001**, *116*, 389.

Mater. Res. Soc. Symp. Proc. Vol. 1038 © 2008 Materials Research Society   1038-O03-03

Polymer Composites for Radiation Detection: Diiodobenzene and Light Emitting Polymer Molecular Solutions for Gamma Detection

Qibing Pei[1], Yongsheng Zhao[2], and Haizheng Zhong[2]

[1]Department of Materials Science and Engineering, University of California, Los Angeles, Engineering V Bldg, Room 3121H, 420 Westwood Plaza, Los Angeles, CA, 90095-1595
[2]Department of Materials Science and Engineering, University of California, Los Angeles, 420 Westwood Plaza, Los Angeles, CA, 90095-1595

ABSTRACT

Conjugated polymers are largely intact by gamma exposure but can be energized in the presence of high-Z compounds. The resulting alteration of the polymer's high optical density and photoluminescence efficiency can be exploited for the detection of gamma radiation with high sensitivity. Diiodobenzene and conjugated polymers mix on the molecular level in solid thin films. Composite films of various thicknesses were conveniently cast from solution and exposed to gamma radiation. The responses of the films to gamma dosage were observed with dramatic changes in ultraviolet-visible absorption and photoluminescence.

INTRODUCTION

Gamma detection is generally based on scintillation or photoelectron generation. Because of the requirement of high-Z for high stopping power, inorganic compounds containing high-Z atoms are often used.[1-3] Synthetic polymers, composed primarily of low-Z carbon, hydrogen and oxygen atoms are not particularly sensitive to gamma exposure. On the other hand conjugated polymers exhibit high photoluminescence efficiency and moderate carrier mobility. They have been studied for the detection of charged particles, such as electrons,[4] protons,[6] and α particles.[7] For high-energy photons such as X and gamma rays, the changes in polymer properties were observed only at high dosages, higher than 1 kGy.[8,9] Borin et al.[10] observed that the sensitivity of poly[2-methoxy-5-(2'-ethylhexyloxy)-p-phenylenevinylene] (MEH-PPV) to gamma photons can be substantially increased when the polymer is dissolved in chloroform. The formation of Cl radicals due to gamma irradiation accelerates the degradation of the polymer. Campbell and Crone[11] showed that the addition of core-shell quantum dots of CdSe-ZnSe increases the scintillation efficiency of MEH-PPV. This method is limited by photoluminescence quenching and scattering loss at high volume fraction of the high-Z quantum dots.

We are exploiting two unique properties of conjugated polymers for gamma radiation: high-sensitivity photoluminescence quenching by charge transfer[12,13] and fluorescence resonance energy transfer[14]. The conjugated polymers are mixed with high-Z compounds on the molecular or nanometer scale. The high-Z compounds absorb gamma photons. The resulting charged species and/or exitons cause dramatic change of the polymer's photophysical properties through charge or energy transfer. Below we report the preparation of polymer-diiodobenzene molecular solutions and their response to gamma radiation.

EXPERIMENT

MEH-PPV and diiodobenzene were first dissolved in chloroform with different weight ratios. Various amounts of diiodobenzene were employed such that the weight ratios of MEH-PPV to diiodobenzene were 1:0 (pure MEH-PPV), 1:20, 1:30, 1:40, and 1:50, respectively. Thin films were prepared by spin coating at 1600 rpm. The composite films were then exposed to a 137Cs gamma ray source (The JLSHEPHERD & ASSOCIATES, Saxon Cs-137 Mark One Irradiator) for different doses. The optical absorption spectra of the films before and after radiation were measured with a Shimazu UV-1700 UV-visible spectrophotometer, and the PL spectra were taken on a Photon Technology International fluorospectrometer. Fourier transform infrared (FT-IR) spectra were measured on films coated on disc-shaped KBr crystals.

RESULTS AND DISCUSSION

Figure 1 displays the optical photographs of the MEH-PPV/diiodobenzene composite films at various weight ratios. Apparently, all films show the characteristic orange color of MEH-PPV with uniform morphology. The amount of diiodobenzene does not influence the film quality or the PL emission color. A slight PL color change in sample E at MEH-PPV:diiodobenzene weight ratio of 1:50 is observed due to phase separation: the freshly prepared films exhibited uniform morphology with intense orange PL emission; but gradually turned slightly opaque with reduced PL intensity.

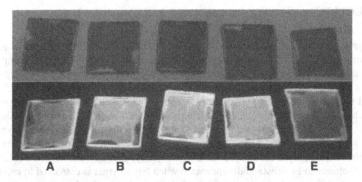

Figure 1. Upper row: Optical photographs of MEH-PPV/diiodobenzene composite films before gamma radiation. The weight ratios of MEH-PPV to diiodobenzene are (A) 1:0; (B) 1:10; (C) 1:20; (D) 1:40; and (E) 1:50. Lower row: corresponding films under UV illumination to show the PL emission.

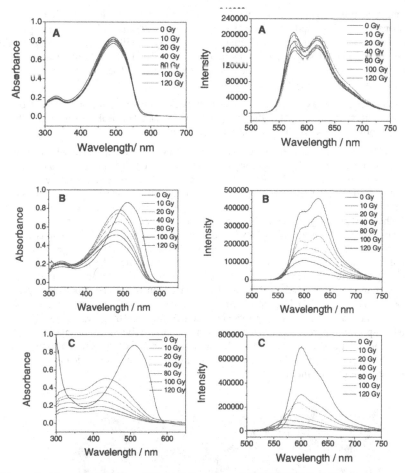

Figure 2. Absorption (left) and PL (right) spectra of MEH-PPV/diiodobenzene composite films with MEH-PPV to diiodobenzene weight ratio being 1:0 (A); 1:20 (B), and 1:50 (C). Insets show gamma dosage (0-120 gray).

Figure 2 shows the absorption and PL spectra of the MEH-PPV/diiodobezene composite films with different weight ratios. The absorption spectra of all films show blue shift of the absorption peak and reduction of peak absorbance with increasing gamma ray dosage. The changes are dependent of diiodobenzene content: pure MEH-PPV shows the least change. At the MEH-PPV to diiodobenzene weight ratio of 1:50, the blue shift of the absorption peak is as much as 90 nm at 10 gray. The gamma exposure also causes substantial PL quenching. Again,

33

diiodobenzene increases the photosensitivity of MEH-PPV to gamma ray. For the MEH-PPV/diiodobenzene (1:50) composite, 120 gray of gamma exposure causes 97% quenching of the PL emission. Figure 3 shows the optical images of the composite films with 1:50 weight ratio of diiodobenzene to MEH-PPV after gamma radiation. At the 80 and 100 grays of dosages, the characteristic MEH-PPV orange color is almost completely bleached. The reduction of the PL intensity is also shown in the images under UV illumination.

Figure 4 shows the peak absorption wavelength, peak absorbance, and peak PL intensity as a function of gamma dosage. Most of the changes are nonlinear. The peak absorbance of the 1:20 composite shows somewhat linear reduction with dosage.

The color change and PL quench are both indicative of degradation of MEH-PPV. The degradation is examined by FTIR spectroscopy as shown in Figure 5. The assignments of the infrared active bands of pristine MEH-PPV are shown in Table 1. There is no obvious change in the IR of pure MEH-PPV after gamma radiation for 100 gray. All characteristic IR-active bands of the polymer, as reported by Scott *et al.*[15] are clearly detected. No carbonyl bands are detectable in the 1700 cm^{-1} region, indicating that the polymer is of good quality and largely intact by gamma radiation. For the MEH-PPV/diiodobenzene composites, two new bands appear below 1700 cm^{-1} and one band appears above 1700 cm^{-1} after 20 gray gamma dosage. The appearance of the band at 1672 cm^{-1} suggests the formation of aromatic aldehydes.[15] The band at 1598 cm^{-1} can be attributed to the asymmetric stretching in the aromatic rings due to the formation of carbonyls.[16] The band at 1737 cm^{-1} has been attributed to the carbonyl absorption of an ester or carboxylic acid. The band at 1737 cm-1 became more predominant with increasing diiodobenzene content and with increasing gamma dosage from 20 to 100 gray. Figure 5 (B) shows the FTIR spectra of pure diiodobenzene which do not reveal significant change of all characteristic vibration peaks of the molecule. Therefore, the spectral changes in the composites are caused by degradation of MEH-PPV. The presence of diiodobenzene dramatically increases the photosensitivity of the polymer to gamma ray.

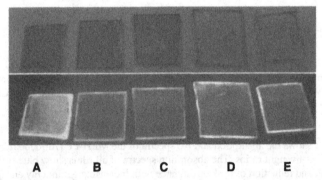

Figure 3. Upper row: Optical photographs of MEH-PPV/diiodobenzene (1:50 by weight) composite film weight after gamma radiation of different doses: (A) 0 gray; (B) 20 gray; (C) 40 gray; (D) 80 gray; (E) 100 gray. Lower row: Corresponding films under UV illumination.

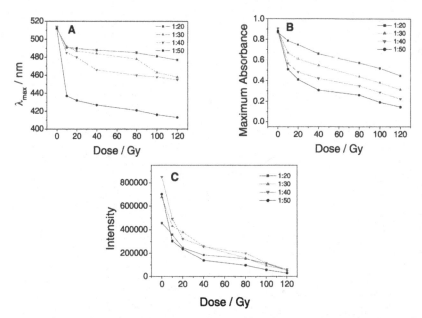

Figure 4. Dose dependence of the absorption wavelength (A), maximum absorbance (B), and PL intensity (C) of MEH-PPV/diiodobenzene composite films. Insets show the weight ratio of MEH-PPV to diiodobenzene.

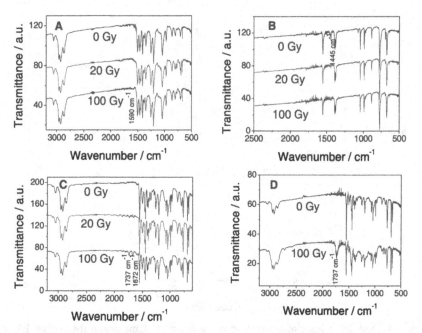

Figure 5. FTIR spectra of thin films of (A) MEH-PPV; (B) diiodobenzene; (C) MEH-PPV/Diiodobenzene (1:10 weight ratio); and (D) MEH-PPV/Diiodobenzene (1:20 weight ratio). The gamma dosages of the films are shown in each chart.

Table 1. IR absorption band assignments for MEH-PPV

Frequency (cm^{-1})	Assignment
2958	CH_3 Asymmetric stretching
2928	C-H stretching
2860	CH_2 stretch
1506	Semicircular phenyl stretch
1464	Anti-symmetric alkyl CH_2
1412	Semicircular phenyl stretch
1380	Symmetric alkyl CH_2 deformation
1204	Phenyl-oxygen stretch
1040	Alkyl-Oxygen stretch
969	Trans double bond CH wag
858	Out-of-plane phenyl CH wag

To investigate whether crosslinking occurs during the degradation of MEH-PPV, the gamma-exposed composite films were immersed in chloroform and sonicated for 1 hour. There were considerable amount of insoluble materials remaining on the substrates, indicating that MEH-PPV is partially crosslinked during gamma radiation.

CONCLUSIONS

The molecular solution of MEH-PPV and diiodobenzene leverages the photophysical properties of the polymer with the high-Z of the additive for enhanced sensitivity to gamma radiation. MEH-PPV is largely intact by gamma ray, but becomes highly sensitive in the presence of diiodobenzene. With increasing diiodobenzene content and higher dosage, the absorption peak blue shifts, peak intensity decreases, and the PL emission intensity declines. Linear degradation with dosage is observed in the composites with 1:20 MEH-PPV to diiodobenzene weight ratio. The degradation was found to cause bleaching and crosslinking of the polymer.

ACKNOWLEDGMENTS

This work is supported by the Defense Threat Reduction Agency, contract #HDTRA1-07-1-0028.

REFERENCES

[1] G. F. Knoll, *Radiation Detection and Measurement, 3rd Edition*, J. Wiley and Sons, New York, **2000**.

[2] A. Owens, *J. Synchrotron Rad.* **2006**, *13*, 143.

[3] A. Owens, A.G. Kozorezov, *Nuclear Instruments and Methods in Physics Research A* **2006**, *563*, 31.

[4] K. Yoshino, S. Hayashi, and Y. Inuishi, *Jpn. J. Appl. Phys., Part 1* **1982**, *21*, L569

[5] H. Kudoh, T. Sasuga, T. Seguchi, and Y. Katsumura, *Polymer* **1996**, *37*, 2903.

[6] K. W. Lee, K. H. Mo, J. W. Jang and C. E. Lee, *Journal of the Korean Physical Society*, **2005**, *47*, 130.

[7] P. Beckerle, H. Ströbele, *Nuclear Instruments and Methods in Physics Research A* **2000**, *449*, 302.

[8] S.C. Graham, R.H. Friend, S. Fung, and S.C. Moratti, *Synthetic Metals* **1997**, *84* , 903.

[9] M. Atreya, S. Li, E. T. Kang, K. G. Neoh, Z. H. Ma, K. L. Tan, W. Huang, *Polymer Degradation and Stability* **1999** *65* 287.

[10] E. A. B. Silva, J. F. Borin, P. Nicolucci, C. F. O. Graeff, T. G. Netto, R. F. Bianchi, *Appl. Phys. Lett.* **2005**, *86*, 131902.

[11] I. H. Campbell, B. K. Crone, *Adv. Mater.* **2006**, *18*, 77.

[12] L. Chen, D. W. McBranch, H. L. Wang, R. Helgeson, F. Wudl, D. G. Whitten, Proceedings of the National Academy of Sciences of the United States of America, **1999**, *96*, 12287.

[13] D. T. McQuade, A. E. Pullen, T. M.Swager, *Chem. Rev.* **2000**, *100*, 2537.

[14] B.W. Van der Meer, G. Coker, S.-Y. Chen, *Resonance Energy Transfer: Theory and Data*, VCH Publishers, Inc., NewYork, 1994.

[15] J. C. Scott, J. K. Kaufman, P. J. Brock, P. R. Di, J. Salem, J. A. Goiltia. *J. Appl. Phys.* **1996**, *79*, 2745.

Mater. Res. Soc. Symp. Proc. Vol. 1038 © 2008 Materials Research Society   1038-O04-02

Crystal Growth and Characterization of CdTe and $Cd_{0.9}Zn_{0.1}Te$ for Nuclear Radiation Detectors

Krishna C. Mandal, Sung H. Kang, Michael Choi, Alket Mertiri, Gary W. Pabst, and Caleb Noblitt
EIC Laboratories, Inc., Norwood, MA, 02062

ABSTRACT

CdTe and $Cd_{0.9}Zn_{0.1}Te$ (CZT) crystals have been studied extensively at EIC Laboratories for various applications including x- and γ-ray imaging and high energy radiation detectors. The crystals were grown from in-house zone refined ultra pure precursor materials using a vertical Bridgman furnace. The growth process has been monitored, controlled and optimized by a computer simulation and modeling program (MASTRAPP). The grown crystals were thoroughly characterized after cutting wafers from the ingots and processing by chemo-mechanical polishing. The infrared (IR) transmission images of the processed CdTe and CZT crystals showed an average Te inclusion size of ~10 μm for CdTe crystals and ~8 μm for CZT crystals. The etch pit density was $\leq 5\times10^4$ cm^{-2} for CdTe and $\leq 3\times10^4$ cm^{-2} for CZT. Various planar and Frisch collar detectors were fabricated and evaluated. From the current-voltage measurements, the electrical resistivity was calculated to be ~ 1.5×10^{10} $\Omega\cdot$cm for CdTe and $2-5\times10^{11}$ $\Omega\cdot$cm for CZT. The Hecht analysis of electron and hole mobility-lifetime products ($\mu\tau_e$ and $\mu\tau_h$) showed $\mu\tau_e=2\times10^{-3}$ cm^2/V ($\mu\tau_h=8\times10^{-5}$ cm^2/V) and $\mu\tau_e=3-6\times10^{-3}$ cm^2/V ($\mu\tau_h=4-6\times10^{-5}$ cm^2/V) for CdTe and CZT, respectively. Final assessments of the detector performance have been carried out using ^{241}Am (60 keV) and ^{137}Cs (662 keV) energy sources and the results are presented in this paper.

INTRODUCTION

CdTe and $Cd_xZn_{1-x}Te$ (cadmium zinc telluride, CZT) are the most attractive materials for room temperature γ-ray and X-ray spectroscopy. Among many candidate materials for γ-ray detectors, CdTe and CZT are the most promising due to their room temperature operation, high average atomic number (Z~50), wide bandgap ($\geq$1.5 eV at 300K) and high density (~5.8 g/cm^3) [1]. Currently used low bandgap Si and Ge detectors can only work efficiently at liquid-nitrogen temperature, which is not suitable for portable room temperature applications. Traditional scintillation detectors connected to photomultiplier tubes are not capable of providing resolutions as high as CdTe and CZT detectors because the energy required for generating one electron-hole pair in scintillator crystals (~50 eV) is much larger than that required for CdTe and CZT (4-5 eV). There have been increasing demands in high-resolution detection and identification of individual isotopes in real-time in various environments, especially for Homeland security applications. In addition, there has been active research in CZT detectors that has shown spectral performance improvement using novel single carrier detector designs such as Frisch collar [2, 3], small pixel [4] and coplanar grid [5]. Due to these advantages, CdTe and CZT have been the primary semiconductor materials for room temperature X-ray and γ-ray detectors in medical imaging, infrared focal plane arrays, national security, environmental monitoring, and space astronomy [6-9].

However, the nuclear spectrometer grade crystal growth of CdTe and CZT by the high pressure Bridgman (HPB) method has suffered from low-yields and small crystal sizes. The HPB growth also has problems including easy defect formation such as Te inclusions, grains and twinning, cracking due to thermal stresses and non-uniform crystal composition caused by zinc segregation. In the crystal growth process, the quality of the as-grown crystal is significantly influenced by complex transport phenomena in the Bridgman system. The melt flow by the buoyancy force significantly affects the solid/melt interface shape and dopant impurity distribution in the as-grown crystal, causing radial and axial segregation that adversely affects the device quality [10]. In addition, the inhomogeneous temperature distribution as well as wall contact can cause mechanical stresses in the crystal and result in high dislocation density. Achieving the dopant uniformity in the grown CdTe and CZT crystals requires precise control of the melt flow and heat and mass transfer in the growth system.

A modified vertical Bridgman growth technique has been used at EIC Laboratories to produce large volume detector grade CdTe and CZT single crystals with a high yield [11 - 13]. In order to reduce defect formation during the crystal growth, we have used numerical modeling and simulation that combines formulation of global heat transfer and thermal elastic stresses. Using this model, the temperature distribution in the furnace and the resulting flow patterns in the melt and the interface shapes were predicted. From these results, thermal stresses in the growing crystal caused by the non-uniform temperature distribution were calculated by using an elastic stress sub-model with special attention to the interaction between the crystal and the ampoule. Based on the simulation results, CdTe and CZT crystals have been grown from in-house zone refined (ZR) precursor materials using a three-zone furnace. The grown crystals (diameter: ~2.5 cm, and length: ~10 cm) have shown promising characteristics for high-resolution room temperature radiation detectors with high resistivity and good charge transport properties. The fabricated detectors have shown very low leakage currents and high-count rates for various radiation sources.

EXPERIMENTAL

Figure 1 shows a schematic of a vertical Bridgman growth furnace along with the temperature distribution in the growth chamber. The Bridgman growth system used for CdTe and CZT growth consists of three zones - a hot zone (~ 1185°C), a cold zone (~ 980°C) and a gradient zone. All the zones of the furnace were independently controlled precisely by temperature controllers and the complete thermal profile was monitored by Adept 2000 (Adept Technology) software. The temperatures in the hot zone and cold zone were specified to create a longitudinal temperature gradient such that the liquid-solid interface is positioned in the adiabatic zone. The ampoule loaded with source materials was dropped from the hot zone to the cold zone at a preset rate. During this process, a series of phenomena took place, including the melting of the source material, solidification of the melt and cooling of the crystal.

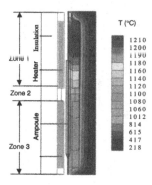

Figure 1. Schematic of EIC's Bridgman growth system and temperature distribution in the growth chamber.

A state-of-the-art computer model, multizone adaptive scheme for transport and phase-change processes (MASTRAPP), was used to model heat and mass transfer in the crystal growth system and to predict the stress distribution of the as-grown crystal [14-15]. In our previous studies [13, 15], numerical simulations have been performed to study the effects of the Grashof number on crystal growth when Marangoni convection was not considered. It was found that the convection was very strong for a Grashof number of 10^6. The flow field got weaker and the interface was less curved when the Grashof number was reduced. In addition, it was shown that thermal stresses caused by the non-uniform temperature distribution as well as the constraint from the ampoule could cause plastic deformation in the growing crystal. From von Mises stress calculations in the crystal, a high stress concentration was noted along the crystal peripheral surface near the interface and the minimum stress could be obtained if there was no contact to the wall [13-15]. In this study, numerical simulations were performed to see the effects of the Marangoni number on fluid flow, temperature distribution and solid-melt interface morphology. Figures 2 (a) and (b) show the effects of the Marangoni number on fluid flow and temperature distribution for $G_r=1\times10^5$. Marangoni convection is a phenomenon of liquid flowing along an interface from places with low surface tension to places with a higher surface tension. The Marangoni number is defined as a ratio of a characteristic diffusion time and a characteristic time for Marangoni driven flow. Figure 2 (c) shows the effects of the Marangoni number on the solid-melt interface morphology. In this figure, we can see that the interface shape changes from partially concave to convex as the Marangoni number increases. The concave interface is ideal because it prevents facet and twin formation. Because the Marangoni number changes by modifying the size of the ampoule and the temperature difference between the hot and cold zones, we can control these parameters to get most desirable interface based on the numerical simulation results. Excellent work has also been carried out by Pandy et al. [16] and by Yeckel et al. [17-19] on the details of numerical modeling and simulation of CZT growth.

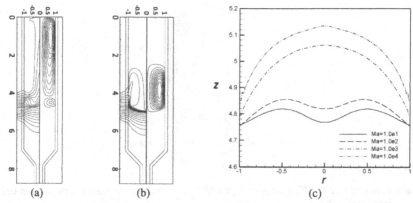

Figure 2. Effects of the Marangoni number on fluid flow and temperature distribution $(G_r=1\times10^5)$ (a) $M_a=1\times10^1$ (b) $M_a=1\times10^3$ (c) Effects of Marangoni number on the solid-melt interface morphology.

For crystal growth, in-house zone refined Cd, Zn and Te precursors were weighed in stoichiometric amounts for CdTe and CZT crystal growth and were vacuum sealed at 10^{-6} torr in a cleaned carbon-coated quartz ampoule ($\geq$3 mm wall thickness). The precursor materials were etched by following the procedures described in our previous papers [13, 15]. The quartz ampoule was sealed by immersing the material containing end in liquid N_2 while sealing the other end. The sealed ampoule was then loaded into the two-zone horizontal furnace for synthesis. The CdTe and CZT polycrystalline charges were prepared by slowly heating the ampoules to a maximum temperature of ~1145°C for CdTe and ~1185°C for CZT. Continuous slow rotation during the synthesis ensured homogeneity. The polycrystalline ingots were then placed in a conically tipped carbon coated quartz ampoule (wall thickness $\geq$3 mm) and sealed under a dynamic vacuum of 10^{-6} torr. The conical tip was used to prevent secondary nucleation and to allow growth with a preferred orientation by holding an oriented CdTe or CZT seed crystal. An axial low temperature gradient (10-15°C/cm at the growth zone) was imposed to stabilize the solid-liquid interface, to suppress evaporation of Cd and Te and to minimize stresses resulting from the thermal expansion coefficients. The sealed ampoule was loaded into the crystal growth furnace and connected to a slow-speed (20 rph) motor. Then, the ampoule was moved downward at a constant speed, resulting in the directional solidification at ~ 0.5 cm/day.

After crystal growth, several CdTe and CZT wafers (up to $10\times10\times10$ mm^3) were cut from the as-grown ingot, ground and lapped (down to 0.3 μm alumina), polished (down to 0.3 μm diamond paste) and etched (1% Br$_2$-methanol dipping for 60s, 0.5% Br$_2$-methanol washing for 30s, and ultra-pure methanol rinsing). The etched wafers appeared bright, shiny and had a good surface finish. Such wafers were used for evaluating physical, optical and electrical properties as well as the detection capabilities of the CdTe and CZT crystals. IR images were taken to characterize Te inclusions/precipitates and their distributions. Scanning electron microscopy (SEM) was used to see the surface defects and morphology and to calculate the etch pit density after Everson etching [20]. Defect analysis was performed by using a two-modulator generalized ellipsometer (2-MGE) [21-22]. Following the material characterizations, detectors were fabricated by depositing a DC sputtered gold contact (50 – 70 A) on one side (rectifying contact) and an e-beamed indium contact (50 – 70 Å) on the opposing side (ohmic contact). Contacts were deposited using a metal mask. Nuclear detection measurements were carried out at room temperature by irradiating CdTe and CZT detectors with various sources including ^{241}Am and ^{137}Cs. The bias to the detector was applied using a Canberra 3102D high voltage power supply. The generated charge signal was amplified with a Princeton Gamma-Tech RG-118 B/C preamplifier, and a Canberra 2022 linear amplifier. It was then fed into a Canberra 31020 multichannel analyzer and evaluated using Genie-2000 software. Figure 3 shows a schematic diagram of the nuclear detection measurement set-up used in this study.

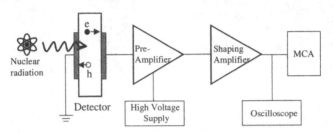

Figure 3. Schematic diagram of the nuclear detection measurement set-up.

RESULTS AND DISCUSSION

Following the growth procedure above, we have been able to grow ≥2.5 cm diameter and ~10 cm long CdTe and CZT ingots. A typical 1 cm^3 CZT single element detector is shown in Figure 4 (a). Energy dispersive spectroscopy (EDS) on four CdTe wafers cut at different axial distances along the grown CdTe ingot showed reasonably stoichiometric uniformity (±0.5% atomic). However, EDS spectra on five axial CZT wafers showed that the Zn distribution profile has deviations of about 5% in stoichiometry at the top and bottom portions of the ingot but less than ±0.1% at the center. Because the Zn distribution coefficient in CdTe [K_{zn} (CdTe) = 1.35] is greater than 1, the axial variation of Zn concentration was higher at the last-to-freeze end of the ingot. However, the Zn concentration radial variation across the sliced wafer was much more uniform (<1%). X-ray photoelectron spectroscopy (XPS) analysis was also performed on CdTe and CZT crystals to determine the chemical composition (Cd/Te and Cd/Zn ratio) in the surface and in the bulk. The results agreed very well with the values obtained from EDS.

After evaluating the stoichiometry of the grown crystals, they were characterized by IR spectroscopy to evaluate Te inclusions. The IR image of a CZT wafer in Figure 4 (b) showed good crystal quality with a small number of Te inclusions (≤10 μm). The IR image of a CdTe wafer showing dark spots (Te inclusions, average diameter ≤20 μm) is presented in Figure 4 (c). An IR image of an improved CdTe crystal grown under Cd-overpressure showing a relatively low concentration and smaller sizes of Te inclusions is shown in Figure 4 (d). Te inclusions adversely affect detection characteristics of CdTe and CZT devices. Thus, the low density of Te inclusions in the grown crystals shows their high potential for a high-resolution nuclear detector. The crystals were also characterized by SEM to evaluate the surface morphologies. The SEM picture in Figure 5 (a) shows very smooth and shiny surfaces without any micro-cracks or major defects. In order to do quantitative analysis of the crystal quality, etch pit density was evaluated by taking SEM picture of the grown single crystal after Everson etching [20]. From Figure 5 (b), the calculated etch pit density was $\leq 3\times 10^4$ cm^{-2}, which also confirms good crystal quality. Further analysis was performed by 2-MGE measurements.

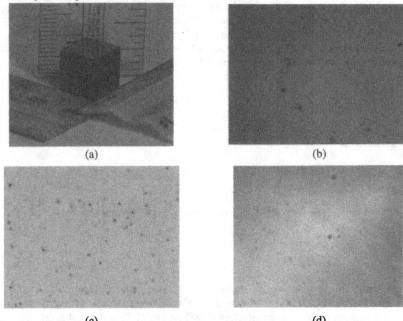

Figure 4. (a) Processed and fabricated single element CZT detector ($10\times 10\times 10$ mm^3); IR images of (b) a CZT and (c) a CdTe wafer; (d) IR image of a CdTe crystal grown under Cd-overpressure showing a relatively low concentration of Te

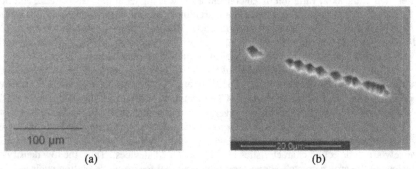

Figure 5. (a) SEM picture of a polished and chemically etched CZT crystal wafer. (b) SEM picture of etch pits revealed after Everson etching.

For 2-MGE measurements, a GaAs PMT was used with a 890 nm laser. The pixel size was 20 μm and the scan was performed on a 3×3 mm² area. The intensity image of Figure 6 (a) showed several small dark regions where much of the 890 nm light is blocked; this is an indication of Te-inclusions and phase separation. Moreover, the images of the retardation and the diattenuation in Figure 6 (b) and (c) showed quite different images, indicating that the phases have different optical properties. Three regions (shown as 1, 2, and 3 on the whole area image) have been selected to zoom. Region 1 showed a distinctive increase in retardation associated with a dark region in the intensity pattern and a slight increase in the diattenuation. This is a strong indication that either the particle is birefringent (Te-inclusions) or the presence of the particle is significantly stressing the surrounding region. Region 2 showed a defect that had very strong signatures in diattenuation, retardation and circular diattenuation, but it was not observable in the transmission intensity. This defect is probably a separate phase, but one that has a band gap greater than 1.39 eV (= 890 nm). Given the large diattenuation, it is hard to explain this defect as a strained portion of the crystal. Region 3 showed a defect that had both a signature in retardation and diattenuation, but was not obvious at all in the intensity. Moreover, the diattenuation and retardation signatures were quite different. It might be a void or bubble defect which created a strain pattern. This may explain the retardation but does nothing for the diattenuation.

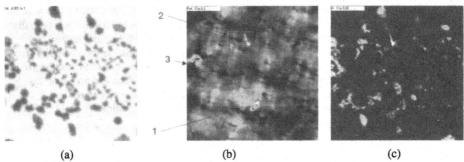

(a)   (b)   (c)

Figure 6. Mapping of a CZT crystal for (a) Transmission intensity (b) Retardation (c) Diattenuation.

After evaluating material properties of the grown crystals, device characteristics were measured after fabricating detectors following the procedures described in the previous section. Figure 7 (a) and (b) show the current-voltage (I-V) characteristics of a CdTe and a CZT detector, respectively. The dimension of the CdTe detector was 9.2×8.6×3.5 mm³ and the electrode area was 8×7 mm². The dimension of the CZT detector was 7×7×5 mm³ and the electrode diameter was 2.6 mm. The bulk electrical resistivity was calculated from the I-V characteristics and was >10¹¹ Ω·cm for CZT and >10¹⁰ Ω·cm for CdTe. The CdTe and CZT detectors showed very low leakage currents at a high bias (below 5 nA at –1000V) due to their high resistivity, which are beneficial for high resolution detectors.

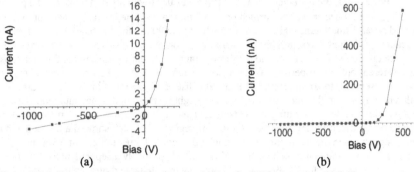

Figure 7. (a) Current-voltage (I-V) characteristics of a CdTe detector; (b) I-V characteristics of a CZT detector.

Based on the I-V characteristics, charge transport properties of electrons and holes were evaluated by measuring pulse positions at different biases. Upon reversal of the polarity of the applied bias, the mobility-lifetime products ($\mu\tau$) for electrons and holes (Figure 8 (left) and (right), respectively) have been extracted using Hecht analysis. An A [241]Am source was used for the $\mu\tau$ product measurement and the results are summarized in the Table I below. It is seen that CdTe has more balanced electron and hole transport properties while CZT has a higher resistivity. By comparing the properties of CZT crystals grown from 6N purity precursors ("as is" from vendor) with that of crystals grown from EIC's ZR precursors, it is clear that the use of zone-refined precursor materials resulted in higher resistivity and electron and hole mobility-lifetime products. Since the CZT crystals grown from 6N purity vendor precursors had a narrower bandgap than those grown from ZR precursors, the bandgap shrinkage could be attributed to the impurities present in the 6N purity precursors.

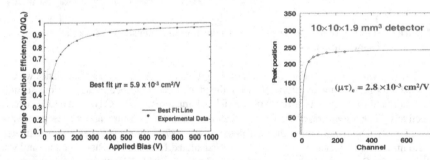

Figure 8. (left) Electron $\mu\tau$ product of a EIC grown CZT sample measured at SMART laboratory, Kansas State University and (right) measured at Brookhaven National Laboratory (BNL), similar results were obtained at EIC Laboratories, Inc.

Table I. Comparison of resistivity and charge transport properties of electrons and holes.

Parameters	CdTe (EIC's ZR Precursors)	$Cd_{0.9}Zn_{0.1}Te$ (6N Purity Precursors from vendor)	$Cd_{0.9}Zn_{0.1}Te$ (EIC's ZR Precursors)
Bandgap [eV, 300K]	1.50	1.52	1.58
Resistivity [Ω·cm]	1.5×10^{10}	$\sim 5 \times 10^{9}$	$\sim 2\text{-}5 \times 10^{11}$
Electron $\mu\tau$ product [cm^2/V]	2×10^{-3}	$\sim 10^{-3}$	$3\text{-}6 \times 10^{-3}$
Hole $\mu\tau$ product [cm^2/V]	8×10^{-5}	2×10^{-5}	$4\text{-}6 \times 10^{-5}$

The digital pulse analysis was performed at BNL and is shown in Figure 9. Figure 9 (a) shows the amplitude on the output signal versus the rise-time and Figure 9 (b) shows the corresponding pulse height spectrum. In Figure 9 (a), the line at amplitude 200 illustrates electronic noise. From the figure, the total energy deposition curve is broad and gets broader with rise time. The broad distribution can be explained by Te inclusions [12 and references therein]. In order to characterize the performance of CdTe and CZT as nuclear spectrometers, nuclear detection measurements were carried out at room temperature by irradiating the CdTe and CZT detectors with [241]Am and ^{137}Cs sources. The pulse height spectra are shown in Figure 10.

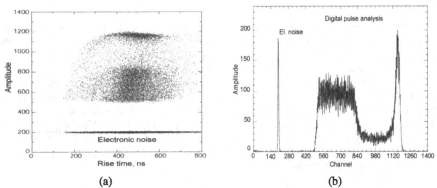

(a)   (b)

Figure 9. (a) Total energy deposition curve for a CZT detector; (b) Corresponding pulse height spectrum. (Data measured at Brookhaven National Laboratory, BNL).

In the pulse height measurements, a negative bias voltage of – 600 V for CdTe and – 1800 V for CZT was applied to the top gold electrode, no polarization effects were found in either case when measured for as long as 72 hours. The leakage current measured was very small (~4-5 nA) as expected from the I-V characteristics. The CdTe detectors clearly detected the 59.6 keV energy of [241]Am with an energy resolution of FWHM = 6.2% (Figure 10 (a)), and the CZT detector detected the 662 keV energy of ^{137}Cs with an energy resolution of FWHM = 17 keV

(2.6%) (Figure 10 (b)). The detection spectra clearly show that the grown single crystals are promising candidates for room temperature solid-state nuclear spectrometers.

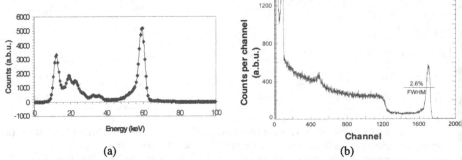

(a)   (b)

Figure 10. (a) Pulse height spectrum measured at EIC on a 6.9×6.9×4.8 mm³ CdTe
detector for ²⁴¹Am source, Bias: - 600V, Gain:10, Shaping time:1.0 μs, Acquisition
time: 120s; FWHM = 6.2%; (b) Pulse-height spectrum measured on a 5×5×10 mm³
CZT detector for ¹³⁷Cs source measured at Brookhaven National Laboratory.

CONCLUSION

Nuclear spectrometer grade CdTe and CZT crystals were grown by a modified vertical
Bridgman growth technique from zone refined precursor materials and characterized by various
physical and opto-electronic methods. The crystal growth conditions were determined by
integrated numerical modeling and simulations using MASTRAPP. As a result, CdTe and CZT
crystals showed high radial uniformity of component concentrations and subsequently improved
optical and electrical properties. Detectors have been fabricated and characterized for charge
transport properties and detection capabilities. Both CdTe and CZT showed desirable charge
transport properties for nuclear spectrometers. The single element detectors fabricated from the
CdTe and CZT crystals have shown energy resolution FWHM 6.2% at 59.6 keV and 2.6% at 662
keV, respectively. Efforts are being directed to improve energy resolution in the range of 60-662
keV by using single carrier detector designs.

ACKNOWLEDGEMENTS

The authors thank Prof. Douglas S. McGregor, Mr. Mark J. Harrison and Mr. Alireza
Kargar of S.M.A.R.T. Laboratory, Kansas State University, Prof. Arnold Burger, Mr. Michael
Groza and Dr. Y. Cui for materials and detector characterization and Prof. Hui Zhang of SUNY,
Stony Brook for simulation and modeling of crystal growth. The authors also thank Drs. Ralph
B. James, Aleksey E. Bolotnikov and Giuseppe Camarda of Brookhaven National Laboratory for
providing IR images and some of the results on pulse height analysis. One of the authors
(K.C.M.) acknowledges partial financial support by the Air Force (Contract Number FA86540-
06-M-5411) and the DOE (Contract Number DE-FG02-07ER84736).

REFERENCES

1. S. U. Egarievwe, K.-T. Chen, A. Burger, R. B. James and M. Lisse, *J. X-ray Sci. and Tech.* 6, 309 (1996).
2. D.S. McGregor, Z. He, H.A. Seifert, D. K. Wehe and R. A. Rojeski, *Appl. Phys. Lett.* 72, 792 (1998).
3. W. J. McNeil, D. S. McGregor, A. E. Bolotnikov, G. W. Wright and R. B. James, *Appl. Phys. Lett.* 84, 1988 (2004).
4. H. H. Barrett, J. D. Eskin, and H. B. Barber, *Phys. Rev. Lett.* 75, 156 (1995).
5. P. N. Luke, *Appl. Phys. Lett.* 65, 2884 (1994).
6. R. B. James, T.E. Schlesinger, J. Lund and M. Schieber, *"Semiconductors for Room Temperature Nuclear Detector Applications"* (Academic Press, New York, 1995).
7. A. Burger, H. Chen, K. Chattopadhyay, J. O. Ndap, S. U. Egarievwe and R.B. James, *SPIE* 3446, 154 (1998).
8. S. Sen, H.L. Hettich, D.R. Rhiger, S. L. Price, M. C. Currie, R.P. Ginn and E. O. McLean, *J. Electron. Mater.* 28, 718 (1999).
9. H. Krawczynski, I. Jung, J. Perkins, A. Burger and M. Groza, *SPIE* 5540, 1 (2004).
10. J.P. Garandet, J.J. Favier and D. Camel, *"Handbook of Crystal Growth"* (Elsevier Science, Amsterdam, 1994).
11. Krishna C. Mandal, S. H. Kang, M. Choi, Alireza Kargar, Mark J. Harrison, Douglas S. McGregor, A. E. Bolotnikov, G. A. Karini, G. C. Camarda, and R. B. James, IEEE Trans. Nucl. Sci. 54, 802 (2007).
12. G. Koley, J. Liu and Krishna C. Mandal, *Appl. Phys. Lett.* 90, 102121 (2007).
13. Krishna C. Mandal, S. H. Kang, M. Choi, J. Wei, L. Zheng, H. Zhang, G. E. Jellison, M. Groza and A. Burger, *J. Electron. Mater.* 36 1013 (2007).
14. H. Zhang, L. L. Zheng, V. Prasad and D. J. Larson, Jr., *J. Heat Transfer* 120, 865 (1998).
15. R.H. Ma, H. Zhang, D.J. Larson, Jr., and Krishna C. Mandal, J. Crystal Growth, 266, 216 (2004).
16. A. Pandy, A. Yeckel, M. Reed, C. Szeles, M. Hainke, G. Müller, and J.J. Derby, J. Crystal Growth, 276, 133 (2005).
17. A. Yeckel, G. Compère, A. Pandy, and J.J. Derby, J. Crystal Growth, 263, 629 (2004).
18. A. Yeckel and J.J. Derby, J. Crystal Growth, 233, 599 (2001).
19. A. Yeckel, F.P. Doty, and J.J. Derby, J. Crystal Growth, 203, 87 (1999).
20. W.J. Everson, C.K. Ard, J.L. Sepich, B.E. Dean, G.T. Neugebauer and H.F. Schaake, *J. Electron. Mater.* 24, 505 (1995).
21. G. E. Jellison, Jr. and F. A. Modine, *Appl. Opt.* 36, 8184 (1997); ibid. 36, 8190 (1997).
22. G. E. Jellison, Jr., C. O. Griffiths, D. E. Holcomb and C. M. Rouleau, *Appl. Opt.* 41, 6555 (2002).

Mater. Res. Soc. Symp. Proc. Vol. 1038 © 2008 Materials Research Society 1038-O04-03

X-ray Topography to Characterize Surface Damage on CdZnTe Crystals

David Black[1], Joseph Woicik[1], Martine C. Duff[2], Douglas B. Hunter[2], Arnold Burger[3], and Michael Groza[3]

[1]NIST, Gaithersburg, MD, 20899
[2]SRNL, Aiken, SC, 29808
[3]Fisk University, Nashville, TN, 37208

ABSTRACT

Synthetic CdZnTe or "CZT" crystals can be used for room temperature detection of α- and γ-radiation. Structural/morphological heterogeneities within CZT, such as twinning, secondary phases (often referred to as inclusions or precipitates), and polycrystallinity can affect detector performance. As part of a broader study using synchrotron radiation techniques to correlate detector performance to microstructure, x-ray topography (XRT) has been used to characterize CZT crystals. We have found that CZT crystals almost always have a variety of residual surface damage, which interferes with our ability to observe the underlying microstructure –for purposes of crystal quality evaluation. Specific structures are identifiable as resulting from fabrication processes and from handling and shipping of sample crystals. Etching was found to remove this damage; however, our studies have shown that the radiation detector performance of the etched surfaces was inferior to the as-polished surface due to higher surface currents which result in more peak tailing and less energy resolution. We have not fully investigated the effects of the various types of inducible damage on radiation detector performance.

INTRODUCTION

Synthetic CdZnTe (CZT) crystals can be used for room temperature detection of gamma radiation. However, the radiation detection properties of CZT crystals vary widely. These variances are not well understood but they have been attributed to structural and morphological heterogeneities within the crystals, such as twinning, secondary phases and polycrystallinity [1-5]. Such heterogeneities often result in poor γ-ray detection performance. To improve our understanding of the relationship between microstructure and detector performance we have developed a broad program to utilize a variety of characterization techniques to investigate the microstructure of CZT crystals and correlate it to detector performance. We report the results of our examination of CZT crystals using synchrotron based x-ray diffraction topographic imaging.

The microstructure of a crystal reveals information about the quality of the crystal. Crystal quality is often attributed to low overall crystal strain, the absence of twins and grain boundaries, and in many cases, the absence of secondary phases [6 and references therein]. Crystal attributes such as these are often good indicators of detector performance [6 and references therein]. These types of features can often be observed by topography and x-ray topography can be used to investigate the crystal quality that is beyond the surface. However, surface damage can obscure the view of the deeper features within crystals that can reveal crystal quality. This paper focuses on the various factors that contribute to surface damage on CZT.

With this information, it is anticipated that unexpected surface damage can be avoided whenever possible so that the microstructural attributes of CZT crystals will be most readily viewed.

METHODS AND MATERIALS

X-ray topography is a well-established, nondestructive characterization technique for imaging the micrometer-sized to centimeter-sized defect microstructure of single-crystals [7-10]. A monochromatic x-ray beam illuminates the single-crystal sample and images of the diffracted beams are recorded. The basis for the technique is that the diffracted intensity from any point on the sample is determined by the local crystal perfection at that point. In other words, the intensity diffracted from an imperfect region of the sample will be different from that from a perfect region. Imperfect regions exist around crystallographic defects, surface and subsurface damage and result from inhomogeneous strain. The x-ray topograph is a map of the spatial distribution of diffracted intensity from the sample surface, which shows the distribution of defects and inhomogeneous strain.

The experimental geometry for recording the surface topograph is shown in figure 1. Topographs can be recorded in transmission through the sample, but for the thick CZT crystals examined here only surface topographs were recorded. A monochromatic and nearly parallel (low divergence) x-ray beam illuminates the sample crystal. A specific set of crystallographic planes is selected and oriented with respect to the incident beam at the Bragg angle according to the equation:

$$E = hc / 2d\sin\theta, \qquad (1)$$

where E is the incident x-ray energy, h is Planck's constant, c the speed of light, d the lattice spacing of the selected set of planes and θ defines the Bragg angle. We used the (333) planes which were oriented parallel to the sample surface. The diffracted beam is viewed in real time on x-ray sensitive video cameras to establish the precise diffraction condition desired and then recorded at 1:1 magnification on high-resolution film. The depth of penetration into the sample was about 9 μm and the sensitivity to local crystallographic orientation is of the order of the rocking curve width of the sample, about 2 arc seconds in this case. The topographs shown here were recorded at the UNICAT/XOR beam line, 33-BM, at the Advanced Photon Source (APS) at Argonne National Laboratory. An incident x-ray energy of 9 keV was selected using a double crystal monochromator with symmetric Si (111) crystals.

The APS x-ray source is partially coherent and this coherence leads to spurious structure in the incident beam from "phase contrast". The basis for phase contrast is the fact that the x-ray beam will refract at discontinuities in the electron density within a material. The refracted beam then propagates in space causing constructive and destructive interference. This interference produces image contrast. One of the advantages of this technique is that low atomic number materials can be examined using relatively high energy x-rays. However, for x-ray topography this contrast mechanism can introduce many spurious features into the incident beam. Any imperfections in the Be windows used to isolate beam line components, or dust on transport beam windows for example produce significant structure in the incident beam. One method to reduce the impact of this incident beam structure is to introduce a random phase object into the beam. This serves to time average the phase contrast, over a time much shorter than any exposure to record a topograph, and therefore smooth it out. In our case a rotating foam disk was placed between the monochromator and the sample. This is successful in eliminating the majority of the phase contrast, but unfortunately does degrade the quality of the beam by adding divergence.

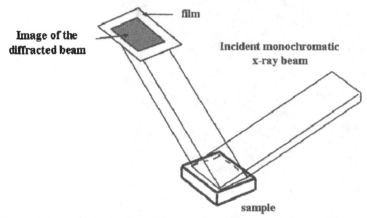

Figure 1. A schematic drawing of the experimental geometry for recording a surface topograph (Adapted from Black and Long, 2004).

The CZT crystals that we examined were grown according to the Modified Vertical Bridgman (MVB) method to have 10% Zn content as described in Li et al. (2001) [11] and procured from Yinnel Tech (South Bend, IN)*. The samples shown here are CZT3-7-2 (12.2 by 11.3 by 3.4 mm^3, polished with 0.05 micron alumina followed by etching with 1% vol. Br:MeOH for 2 minutes), CZT3-7-7 (11.1 by 11.7 by 11.7 mm^3) polished with 0.05 micron alumina followed by gentle polish with BrMeOH on soft felt pad) and sample YT#2, (half etched in Br:MeOH and polished with 0.05 micron alumina and half with polish with 0.05 micron alumina and no etch). All samples were grown by MVB and were detector grade materials.

RESULTS AND DISCUSSION

An example of the most common types of surface damage that we have observed using topography is shown in figure 2. The predominant structure that can be observed throughout the sample surface is residual fabrication damage, which is typical of the type of polishing process that uses abrasive grit on a rotating wheel (where the abrasive particles move in long arcs across the sample surface). This form of surface damage is typical of that induced when a polishing wheel is used to polish a surface. Critical to successful polishing is the removal of sufficient material at each step so that the damage from the preceding grit is eliminated. As the grit size is reduced, this polishing structure become finer and finer until it is no longer visible when examined in an optical microscope at magnifications typically about 100X. Under a microscope,

Figure 2. A symmetric (333) topograph showing several types of surface damage observed on CZT crystals. Fabrication damage is seen over the entire surface as the randomly oriented, long and dark arcs. Shipping damage is identified by the series of features indicated by the double arrow on the left side of the image. The enlarged inset shows it's typical "puff ball" structure. The circular feature at the bottom right corner is from the application of a magic marker "dot" on the surface.

the sample surface may appear smooth, mirror-like and damage free. However, underlying residual damage could still exist. The action of the abrasive particles disrupts the sample surface during the process of removing material and smoothing out the surface. An example of underlying damage is shown schematically in figure 3. Without the removal of additional surface material through subsequent polishing steps, the "legacy" damage will remain on an otherwise highly polished surface. This damage is characterized by local disruptions of the long-range order of the crystal, and x-ray topography is most sensitive to detecting this type of damage. In many cases, topography often can image the types of damage that are not observable by optical techniques, an example is shown in figure 4.

A second type of damage that we observed is shown in figure 2. We refer to this as "shipping" damage. This type of damage occurs when the sample moves within its packing case during shipping and rubs against the packing material. Its two primary characteristics are that it is localized in small areas (i.e., the contact points with the packing material), and that the structure within these areas is random. This feature is "puff ball-like". It can often occur along a line as indicated in figure 2.

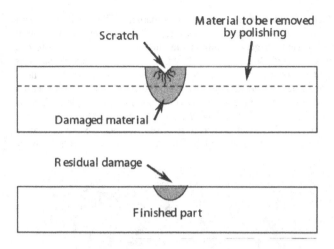

Figure 3. A schematic drawing of how damage below the surface from abrasive polishing can remain even when the surface is flat and smooth.

The last type of damage that we observed (shown figure 2) is from contact with a writing instrument such as a "magic marker". Such a feature is the result of applying a felt tip marker "dot" on to the sample surface to indicate physical orientation and to identify which surface has been examined. Our topography studies rely on known physical orientations and well defined growth planes. Therefore, the marking of a surface on CZT cubes or wafers (to retain its

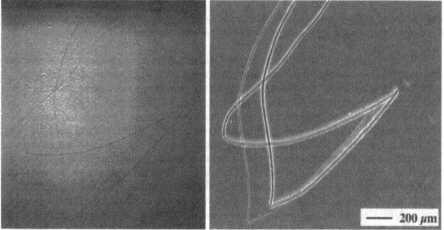

Figure 4. A comparison between the images of a scratch on a sapphire sample in the optical microscope, on the left, and from the (0006) x-ray topograph on the right. The secondary feature observed in the topograph is not visible in the optical image.

physical orientation) prior to topographic studies is almost unavoidable. CZT is sufficiently soft so that the abrasive effect of the marker tip causes surface damage. For example, the cleaning of a sample with alcohol or other solvent by gently rubbing the surface with a cotton swab can result in the type of damage on a polished CZT surface is shown in figure 5. Although care was taken to rub lightly, the applied force of the cotton swab was sufficient to "scratch" the surface, despite the absence of visible scratches on the sample. Another procedure that is often used on CZT crystals, and most other crystals as well, is to use tweezers to manipulate samples. There is an obvious need to avoid direct hand or finger contact with a crystal during handling. However, a highly localized force from the small contact area causes severe damage to the crystal as indicated in figure 5. This damage is quite deep and has been observed on virtually every CZT sample.

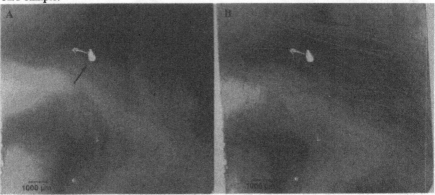

Figure 5. Before (A) and after (B) images of a CZT surface that was cleaned with a cotton swab and alcohol. The feature indicated by the arrow is "tweezer" damage.

Surface damage, and in particular the residual fabrication damage that we often observed in this study, can obscure the underlying microstructure of interest. One method to eliminate surface damage is to etch the sample to chemically remove material. For CZT an effective etch is a 1%Br:MeOH solution which will remove material at about 3 μm/min. As shown in figure 6, etching the surface for 2 minutes removed enough material to eliminate the fabrication damage and permit the viewing of the underlying microstructural details. Microstructural details such as dislocations and subgrain structures are now easily observed as in figure 6. However, contrary to expectations, we have found that the electrical properties and the radiation detector performance of the etched surfaces are inferior to that of the as-polished surfaces [12, in press]. To study detector performance, Au contacts are normally applied to the surface of CZT detectors after polishing with 0.05 μm alumina to remove artifacts from crystal sectioning and after chemical etching with Br:MeOH to remove residual mechanical surface damage. However, etching results in a Te-rich surface layer that is prone to oxidize. Our studies show that CZT surfaces that are only polished (as opposed to polished and etched) can be contacted with Au and will yield lower surface currents. Due to their decreased dark currents, these as-polished surfaces can be used in the fabrication of gamma detectors that exhibit a higher performance than polished and etched surfaces with relatively less peak tailing and greater energy resolution.

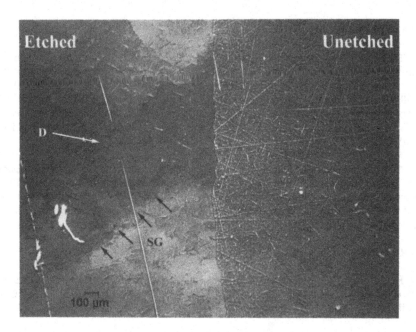

Figure 6. A demonstration of the effectiveness of etching to remove surface damage. In the unetched condition, the underlying microstructure cannot be resolved. When the fabrication damage is removed by etching, the microstructural details such as dislocations (D) and subgrain boundaries (SG) become visible in the topograph. The two white lines on the etched side are from damage caused after etching.

Based on the findings from our study, we recommend using care when handling soft materials such as CZT to avoid unknowingly introducing physical damage to the crystal when handling. We have not fully evaluated the effects that the various types of induced surface damage have on radiation detector performance or other factors related to detector performance (such as resistivity and electron mobility). However, it is important to be aware that handling can induce effects that are not readily observed with the human eye or a light microscope. Finally, the selection of the sample containment that will be used for the transportation of these types of materials should be designed to help minimize the effects of shipping damage.

CONCLUSIONS

X-ray topography is extremely sensitive to localized damage at and near the surface of single-crystals. Using this tool, we have observed pervasive surface damage on CZT crystals. The damage has unique identifiable features and is ascribed to several different mechanisms including fabrication, shipping and handling, cleaning of the surface with solvents and from manipulating the crystals using tweezers. Etching has been shown to remove the residual surface damage, but leaves a surface less suitable for application as a gamma ray detector.

ACKNOWLEDGMENTS

This work is supported by the National Nuclear Security Administration's Office of Research and Development (NA-22). Use of the APS was supported by the U.S. Department of Energy, Office of Science, Office of Basic Energy Sciences, under Contract No. W-31-109-ENG-38. The authors at Fisk University gratefully acknowledge financial support from NSF for Center of Research Excellence in Science and Technology (CREST) grant no. 0420516, and from US DOE Office of Nonproliferation Research and Development (NA-22), Grant No. DE-FG52-05NA27035.

* Certain commercial equipment, instruments, or materials are identified in this paper in order to specify the experimental procedure adequately. Such identification is not intended to imply recommendation or endorsement by the National Institute of Standards and Technology, nor is it intended to imply that the materials or equipment identified are necessarily the best available for the purpose.

REFERENCES

1. J. R. Heffelfinger, D. L. Medlin and R. B. James in *Semiconductors for Room-Temperature Radiation Detector Applications II*, Edited by R.B. James, T.E. Schlesinger, P. Siffert, M. Cuzin, M. Squillante, W. Dusi, M. O'Connell (Mater. Res. Soc. Symp. Proc. **487**, Warrendale, PA, 1998) pp. 33-38.

2. M. Schieber, T. E. Schlesinger, R. B. James, H. Hermon, H. Yoon and M. Goorsky. J. Crystal Growth 237-239, 2082 (2002).

3. C. Szeles and M. C. Driver. in *Hard X-Ray and Gamma-Ray Detector Physics and Applications,* Edited by F. P. Doty and Richard B. Hoover (Proceedings of SPIE Volume: 3446, 1998) pp. 1-8.

4. C. Szeles, S. E. Cameron, J-O. Ndap and W. C. Chalmers., IEEE Trans. Nuclear Sci. 49, 2535 (2002).

5. J. Shen, D. K. Aidun, L. Regel and W. R. Wilcox., J. Crystal Growth 132, 250 (1993).

6. T. E. Schlesinger, J. E. Toney, H. Yoon, E. Y. Lee, B. A. Brunett, L. Franks, R. B. James. Mater. Sci. Engin. 32, 103 (2001).

7. B. K. Tanner, *X-ray Diffraction Topography* (Pergammon, Oxford, 1976)

8. B. K. Tanner in *X-ray and Neutron Dynamical Diffraction Theory and Applications.* Edited by A. Authier, S. Lagomarsino and B. K. Tanner (Plenum Press, New York, 1996)

9. B. K. Bowen and B. Tanner in *High Resolution X-ray Diffractometry and Topography* (Taylor and Francis, London 1998)

10. D. R. Black and G. G. Long, "X-ray Topography", NIST SP960-10 (2004).

11. L. Li, F. Lu, K. Shah, M. Squillante, L. Cirinano, W. Yao, R. W. Olson, P. Luke, Y. Nemirovsky, A. Burger, G. Wright, R. B. James, Nuclear Science Symposium Conference Record, 2001 IEEE 4, 2396 (2001).

12. M. C. Duff, D. B. Hunter, A. Burger, M. Groza, V. Buliga, and D. R. Black. "Influence of surface preparation technique on the radiation detector performance of CdZnTe," *Applied Surface Science*
(in press).

Mater. Res. Soc. Symp. Proc. Vol. 1038 © 2008 Materials Research Society   1038-O05-06

Influence of Gamma-Irradiation on Dielectric Properties of Recycled Polyethylene Composites

Ulmas Gafurov, and Zekie Fazilova

Composite Materials, Institute of Nuclear Physics, Ulugbek, Tashkent, INP, Tashkent, 702132, Uzbekistan

Abstract

The influence of γ-irradiation on the dielectric losses (tg δ) and permittivity (ε') of recycled high density polyethylene (HDPE^R) thermoplastic dynamic vulcanizates (TDV) has been investigated. The materials and composites that were studied include   HDPE^R:EPDM, HDPE^R+EPDM+GTR, HDPE^R+EPDM+GTR/plast; EPDM (ethylene-propylene diamine).

The experimental data obtained shows significant differences between dielectric properties different samples, as well as between the unirradiated and γ-irradiated samples.

Keywords: dielectric losses(relaxation), dielectric permittivity; polyethylene; polymer composites, thermoplastic dynamic vulcanizates (TDV); γ-irradiation

Introduction

The most dynamic developments in the microelectronics industry are based on the efficient utilization of radiation-crosslinkable negative photoresist polymers and the radiation-degradable positive photoresist polymers. Rapid prototyping and rapid tooling have become indispensable methods in the continuously renewing manufacturing technologies of metal and plastic parts for almost all the industrial branches.

In general, the irradiation of polymers yields a similar pathway of development in the early phases, mostly in the radiation cross-linking of polymers and in the radiation sterilization of polymer products [see for example refs. 1-7].

It is known that widely used thermoplastic elastomers (TPE) can be prepared on the base of mixtures composed of non-vulcanized raw rubber and thermoplastic polymers like polyolefins, see refs. [8,9].   Their properties can be notably improved via application of dynamic vulcanization technology and some types of radiation [10-11]. Vulcanization can be carried out chemically via addition of various curing agents or by means of various types of radiation [12]. Phase modification occurs upon controlled irradiation (e. g. gamma- or electronic), namely: cross-linking, decomposition, changing of crystallinity, chemical bonding between the phases, etc. [10-13].

Properties of the TPE materials obtained can be tuned by controlling the ratio of the thermoplastic and rubber-like components. The chemical interconnection between the surface of a binder and the PE matrix was investigated [14] by means of irradiation combined with different types of chemical treatment of the binder, in order to improve the strength properties of glass fiber reinforced plastics. The investigations of the complex action of surface treatment of GFRPs (Glass Fiber Reinforced Plastics) based on thermoplastic matrices and radiation treatment on the strength properties of model glass fiber reinforced plastic, made it possible not only to reveal a modifying effect of radiation, but also to assess the radiation stability of the GFRPs. Fainleib et al. [15] established that irradiation of recycled polypropylene, PP^R, leads to an essential improvement of the mechanical characteristics of TPE prepared on its base.

In the case of the insulating polymeric materials, such as the polyethylenes, polypropylenes and their composites, it is of essential interest to understand the correlations between the structural changes and the dielectric properties. Ref. [16] reports on the dielectric behavior of different polyethylenes irradiated to different absorbed doses of gamma radiation. The dielectric relaxation behavior is related to the changes in the initial structure of different polyethylenes and to the radiation-induced processes of oxidative degradation and cross-linking. The point where a sharp change in the dielectric properties of the material was observed upon ionization processing using fixed intensity partial discharges was selected as the structurally sensitive parameter in this study.

The influence of γ-irradiation on the dielectric properties (dielectric loss and permittivity) of recycled high density polyethylene thermoplastic composites – TDV's (thermo-dynamic vulcanizates) has been investigated.

Experimental

The dielectric losses and permittivity was measured using an E8-4 bridge at a frequency of 1 kHz.
The testing temperature was varied from −150 to 120°C at the heating and cooling rate of 2 grad/min.

The samples were γ-irradiated with different doses using a ^{60}Co-source with an energy of 1.25 MeV. The following materials were studied (EPDM =ethylene-propylene diamine) :

Sample number	Composition	Component content, wt. %
1	HDPE^R / EPDM	40/35
2	HDPE^R / EPDM / GTR	40/35/25
3	HDPE^R / EPDM/ (GTR/Plast)	40/35/25(1/1)

The samples were obtained from the Institute of Macromolecular Chemistry of the National Academy of Sciences of Ukraine (Prof. Alexander Fainleib).
Gamma-irradiation was carried out in the channel of the gamma-source of the Institute of Nuclear Physics of A.S. of Uzbekistan (^{60}Co, energy 1,25 MeV).

Immediately after a period of irradiation a dielectric measurement of the sample was measured and then it was again placed in the channel for the further irradiation. Fourier Transform infrared (FTIR) and attenuated total reflection (ATR) spectra were registered by means of a Bruker IFS-88 spectrophotometer.

The measurements and analysis show that the character of the dielectric pattern changes was qualitatively similar in all samples and so we shall show the typical dielectric parameters changes on an example of some vulcanizates on the basis of the recycled polyethyene.

Results and Discussion

The experimental data obtained are shown   for the composite HDPE^R:EPDM:GTR   in Figure 1 below.

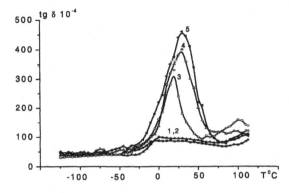

Figure 1. Dielectric relaxation (dielectric losses) of TDV (HDPER:EPDM:GTR) vs. temperature and gamma- irradiation dose. Doses of gamma- irradiation: 0kGy(1), 10kGy(2); 20kGy(3); 50kGy(4); 100kGy(5).

One can see the significant difference between the unirradiated and γ-irradiated samples. The figure 1 shows that dielectric losses curve for γ-irradiated vulcanizate is characterized by presence of sharp peak at the temperature of about 20°C (for dose 20 kGy), and the peak is shifted to higher temperatures with increasing dose. The mentioned peak is assigned to association of absorbed water molecules with polar groups of polymer. It is confirmed by the measurements of the relaxation in regimes of temperature increasing and decreasing.

At the further irradiation (up to 2000 kGy) this losses peak is shifted towards the high temperature region both for vulcanizate HDPE[R]:EPDM and for vulcanizate HDPE[R]:EPDM:GTR.
It is connected with increase in polar group content [17] and, as consequence, with the strengthening of interchain interaction . The data of IR (FTIR) measurements show that gamma - irradiation stimulates growth of polar oxidated groups number (see below and also Figure 2).

For TDV - HDPE[R]:EPDM:GTR/plast (sample 3) the losses intensity raising (growth of the tg δ maximum) is occurred only at doses above 500kGy and the peak shifts to the higher temperature region. That also can be connected with formation of unstable peroxide radicals and the subsequent oxidized groups increasing.

The data of IR (FTIR) spectra shows the carbonyl groups number raising. The IR spectra for various samples had a similar view and we shows it on figure 2 for the vulcanizate HDPE[R]:EPDM:GTR/plast. The IR absorption intensity growth in IR spectra region at 1715см $^{-1}$ shows the carbonyl C=O (ketones and aldehydes) groups number increasing.

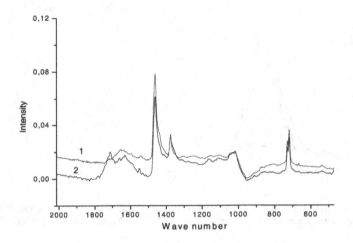

Figure 2. IR (FTIR) spectrum of TD vulcanizate HDPE^R:EPDM:GTR/plast-2 (sample 3) before (1) and after gamma - irradiation with dose ~1200 kGy (2).

Dose dependence data of permittivity for vulcanizate HDPE^R:EPDM (sample 1) is showed in Figure 3.

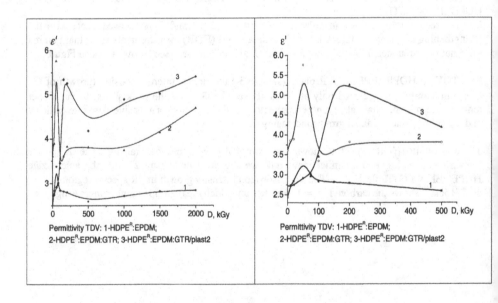

Figure 3. Permittivity TDV: 1-HDPE^R:EPDM; 2-HDPE^R:EPDM:GTR; 3-HDPE^R:EPDM:GTR/plast.

Under irradiation up to dose of 50 kGy the permittivity raising is observed; it may be connected not only to oxidizing processes but also to the presence of peroxides - long-lived products of radiolysis of composites [18].
At the further irradiation up to dose of 100 kGy the permittivity diminution for TDV - HDPER:EPDM and HDPER:EPDM:GTR was pointed out, that is the consequence of radiation-stimulated cross-linking formation [19-22].

For TDV containing the plasticized filler (PPR+EPDM+GTR/plast) the growth of permittivity has pointed out up to dose of 200 kGy, and with increasing in dose up to 500 kGy there is some diminution of ε' what can be a consequence of crystallinity decreasing of polymeric basis of the TDV composite and plastification by dectruction products [21,23,24].

The permittivity growth of the TDV's - $HDPE^R$:EPDM and $HDPE^R$:EPDM:GTR above 500 kGy is observed.
The density is enhanced as a result of irradiation stimulated vulcanizates crosslinking. The permittivity growth with irradiation dose raising is more than it is expected starting from density increasing. Probablly, ε' increases at irradiation not only owing to density growth, but also owing to occurrence of unsaturation.

From the obtained data it is possible to note the permittivity stability for the sample 1 (HDPER:EPDM) above 100 kGy up to high gamma – irradiation doses and for the a sample 2 ($HDPE^R$:EPDM:GTR) at irradiation doses of 100-1000 kGy. And for the plasticized sample 3 (HDPER:EPDM:GTR/plast) at doses higher 200 kGy and for TDV- $HDPE^R$:EPDM:GTR above 1000 kGy it is possible to say only about relative stability when occurs some growth of the permittivity value though it is slow.

Conclusions

The experimental data obtained shows significant differences between dielectric properties different samples, as well as between the unirradiated and γ-irradiated samples. The dielectric losses curve of γ-irradiated vulcanizates are characterized by presence of sharp peak at the temperature of about 20°C (for dose 20 kGy), and the peak is shifted to higher temperatures with increasing dose.
The mentioned peak is assigned to association of absorbed water molecules with polar groups of polymer. It is confirmed by the measurements of the relaxation in regimes of temperature increasing and decreasing.
At the further irradiation (up to 2000 kGy) this losses peak is shifted towards the high temperature region both for vulcanizate $HDPE^R$:EPDM and for vulcanizate $HDPE^R$:EPDM:GTR. It is connected with increase in polar group content (data of IR (FTIR) measurements) and, as consequence, with the strengthening of interchain interaction.
For TDV - $HDPE^R$:EPDM:GTR/plast (sample 3) the losses intensity raising (growth of the tg δ maximum) is occurred only at doses above 500kGy and the peak shifts to the higher temperature region. That also can be connected with formation of unstable peroxide radicals and the subsequent oxidized groups increasing.
Under irradiation up to dose of 50 kGy the permittivity raising is observed; it may be connected not only to oxidizing processes but also to the presence of peroxides - long-lived products of radiolysis of composites. At the further irradiation up to dose of 100 kGy the permittivity diminution for TDV - HDPER:EPDM and HDPER:EPDM:GTR was pointed

out, that is the consequence of radiation-stimulated cross-linking formation.
The permittivity growth of the TDV's - HDPE^R:EPDM and HDPE^R:EPDM:GTR above 500 kGy is observed.

From the obtained data it is possible to note the permittivity stability for the sample 49-III (HDPER:EPDM) above 100 kGy up to high gamma – irradiation doses and for the a sample 2 (HDPE^R:EPDM:GTR) at irradiation doses of 100-1000 kGy. And for the plasticized sample 3 (HDPER:EPDM:GTR/plast) at doses higher 200 kGy and for TDV-HDPE^R:EPDM:GTR above 1000 kGy it is possible to say only about relative stability when occurs some growth of the permittivity value though it is slow.

References
1. The Effects of Radiation on High Technology Polymers, (ACS Symp. Ser. 381). (Eds. E. Reichmanis, J.H. O' Donnell), American Chemical Society, Washington, DC, 1989.
2. Radiation Curing of Polymers. (Ed. D.R. Randell), Royal Society of Chemistry, London, 1988.
3. Guven, O., Cross-linking and Scission of Polymers. (NATO ASI Ser. C, 292). Kluwer Academic Publ., Dordrecht, 1990.
4. Radiation Engineering (Ed. G.G. Eihholz), Marcel Dekker, New York, 1972.
5. Bradley R., Radiation Technology Handbook. Marcel Dekker, New York, 1984.
6. Woods R. J., Pikaev A.K., Applied Radiation Chemistry. Radiation Processing. Wiley, New York, 1994.
7. Czvikovszky T., Radiation Physics and Chemistry, vol. 67, 2003, pp. 437–440.
8. Adhikari B, De D, Maiti S., Prog. Reclamation and recycling of waste rubber. Prog. Polym. Sci. 2000; 25: 909- 948.
9. Nevatia P, Banerjee TS, Dutta B, Jha A, Naskar AK, Bhowmick AK., J. Appl. Polym. Sci., vol. 83, 2002, p.:2035- 2042.
10. Akiba M., Hashim AS.,Prog. Polym. Sci., vol. 22, 1997, pp. 475-521.
11. Karger-Kocsis J., Thermoplastic rubbers via dynamic vulcanization, Ch. 5 in "Polymer blends and Alloys" (Eds.:G.O.Shonaike and G.P.Simon), Marcel Dekker, N.Y., 1999, pp.125-153.
12. Naskar AK, Bhowmick AK, De SK., Thermoplastic Elastomeric Composition based on Ground Rubber Tire. Polym Eng Sci 2001; 41 (6):1087-1098.
13. Bhattacharya A., Radiation and industrial polymers. Prog. Polym. Sci. 2000; 25:371-401.
14. Smirnov Y N, Allayarov S R, Novikova E V, Belov G P, Barelko V V ,Int. Polym. Sci. and Technol., vol. 32, no. 4, 2005 (Article translated from Plasticheskie Massy, No.9, 2004, pp.8-10)
15. Fainleib A.M., Tolstov A. L., Grigoryeva O.P., Starostenko O.N., Danilenko I.Yu., Bardash L. V., Gafurov U.G.,, "Method of producing of thermoplastic elastomer based on recycled polypropylene", Ukraine Patent Application № a 2005 01764.
16. Suljovrujic Edin, Some Aspects of Structural Electrophysics of Irradiated Polyethylenes, IRAP 2004.
17. Gedde U.W., Liu F., Hult A., Sahlen F. and Boyd R.H., Polymer, vol.35, pp. 2056-2063.
18. Frelih G., Theory of dielectrics.// Moscow: Inostrannaya literatura, 1960 (Russ).
19. Electrical properties of polymers.Ed.by B.I.Sazhin.// L., 1970. Last publication-1977 (Russ).
20. Singh A, Radiat. Phys. Chem.,vol. 60, 2001, pp.453-459.
21. Sauer B.B., "Dielectric Relaxation in Polymers: Molecular Mechnismus, Structure-Property Relationship, and Effects of Crystallinity", in: Performance in Plastics, ed. by Witold Brostow, Hanser Publishers, Munich 2001, pp. 208-237.

22. Suarez J.C.M., Mano E.B., "Characterization of degradation on gamma-irradiated recycled polyethylene blends by SEM", Polymer Degradation and Stability, 75, (2001), pp.217-225.
23. Briskman B.A., Milinchuk V.K.,Chemistry of high energies,vol.23,№ 3,1989, pp.195-206 (Russ.).
24. Makhlis F.A., Radiation chemistry of elastomers. Moscow: Atomizdat, 1976, 221p. (Russ.)

This work was supported by the EC (STCU Project № 3009).

Mater. Res. Soc. Symp. Proc. Vol. 1038 © 2008 Materials Research Society 1038-O05-07

Unifying Chemical Bonding Models for Boranes

Mao-Hua Du[1], Susumu Saito[2], and S. B. Zhang[3]

[1]Materials Science & Technology Division and Center for Radiation Detection Materials and Systems, Oak Ridge National Laboratory, PO Box 2008 MS6114, Oak Ridge, TN, 37831

[2]Department of Physics, Tokyo Institute of Technology, 2-12-1 Oh-okayama, Meguro-ku, Tokyo, 152-8551, Japan

[3]Department of Physics, Applied Physics, and Astronomy, Rensselaer Polytechnic Institute, Troy, NY, 12180

ABSTRACT

We demonstrate, based on first-principles calculations, that chemical bonding in deltahedral boron hydrides, $B_nH_n^{2-}$, also known as *closo* boranes, can be understood within the three-center two-electron (3c2e) bonding model in line with other families of boranes. We show that bonding in the triangular lattice of $B_nH_n^{2-}$ cages can be described by delocalized *resonant* 3c2e bonding. We also find that the reason for all the $B_nH_n^{2-}$ to be dianions can be attributed to the reduction of boron coordination number in the deltahedral cage structure from that of boron sheet with triangular lattice.

INTRODUCTION

The nature of chemical bonding is a fundamental problem in chemistry and physics. Lewis's seminal work laid out the general concept of bonding based on shared electron pairs between atoms,[1] which has been later applied to a wide range of molecules, clusters, and extended systems. The most commonly observed bond in a molecule is the localized two-center two-electron (2c2e) bond, formed by sharing two electrons between two neighboring atoms. However, many molecules may also have multi-center bonds. Boron hydrides, or boranes, belong to this class of molecules by exhibiting a mixture of three-center two-electron (3c2e) and 2c2e bonds.

Boron has a rich family of hydrides, such as *closo*, *nido*, and *arachno* boranes.[2, 3, 4] The *closo* boranes ($B_nH_n^{2-}$) have cage structures, which are most stable family of boranes. The *nido* (B_nH_{n+4}) and *arachno* (B_nH_{n+6}) boranes have open structures with their skeletons conveniently seen as *closo* boranes with one and two vertices absent, respectively.[5] Carbon is the only other element to also form a complex series of hydrides, i.e., hydrocarbons. However, chemical bonding in boranes is qualitatively different from that in hydrocarbons. For instance, the simplest borane, B_2H_6, has the same stoichiometry as C_2H_6 but with two fewer electrons. Because the 2c2e σ bonding mechanism perfectly explains the structure of C_2H_6, it naturally cannot explain the stability of B_2H_6. To overcome this problem, the concept of three-center two-electron bonding was developed by Lipscomb and co-workers.[6] In the 3c2e picture, the molecule consists of atoms arranged in a triangular network. Each of the three atoms at the vertices of the triangular structure contributes one atomic orbital (AO) such that the three AOs hybridize to form *one* doubly-occupied bonding molecular orbital (MO) and *two* empty antibonding MOs. The chemical bonding of B_2H_6 can thus be explained as having two 3c2e B-H-B bonds and four 2c2e B-H bonds, as illustrated in Fig. 1(a). A combination of the 3c2e and 2c2e bonds can explain the bonding of many boranes, e.g., the families of *nido* and *arachno* boranes.[7] Such approach is often called Lipscomb's topological model. However, this model does not explain the chemical bonding of *closo* boranes, $B_nH_n^{2-}$, because no spatial allocation of the 3c2e and 2c2e bonds can match the symmetry of the *closo* boranes.[8]

The *closo* boranes have the deltahedral structures and are stable as dianions.[1] Among them, the $B_{12}H_{12}^{2-}$ cluster with icosahedral symmetry [Fig. 1(b)] is the most stable. A *closo* borane ($B_nH_n^{2-}$) has $4n + 2$ valence electrons and $5n$ valence atomic orbitals (AOs), among which $2n$ electrons participate in the 2c2e B-H bonding (using $2n$ AOs) and the remaining $2n + 2$ electrons participate in the skeletal bonding of the boron cage (using $3n$ AOs). The requirement of having $2n + 2$ skeletal electrons or $(2n + 2)/2 = n + 1$ skeletal bonds is known as the Wade's rule.[9] Wade's rule has also been generalized to *nido* and *arachno* boranes.

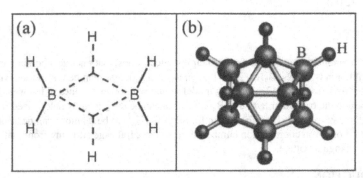

Figure 1. (a) A schematic B_2H_6. Dashed lines indicate the two 3c2e bonds. (b) The calculated structure of icosahedral $B_{12}H_{12}^{2-}$.

According to Wade's rule, the number of skeletal orbitals for *closo* ($B_nH_n^{2-}$), *nido* (B_nH_{n+4}), and *arachno* (B_nH_{n+6}) boranes are $n + 1$, $n + 2$, and $n + 3$, respectively. Wade's rule is an empirical electron counting rule, based on the recognition of the structure relationship between the *closo*, *nido*, and *arachno* boranes[5] and generalized from the earlier molecular orbital (MO) calculations[10, 11, 12]. In fact, with the knowledge that there are n 2c2e B-H bonds in any *closo* ($B_nH_n^{2-}$), *nido* (B_nH_{n+4}), and *arachno* (B_nH_{n+6}) boranes, Wade's rule is just a mathematical consequence of orbital counting (the number of total bonding orbitals minus the number of 2c2e B-H orbitals). It is also straightforward to prove the Wade's rule by checking how many bonding MOs are needed to form an electronic close shell using numerical calculations. Early MO calculations[11] and even the graph-theoretical approach[13] (a mathematical analog to the Hückel theory using a graphical description of the interatomic interactions) can give the result of $n + 1$ bonding orbitals for the skeletal bonding in the *closo* boranes. (The history of the computational studies of boranes has been reviewed recently.[2]) However, the numerical calculations do not provide an explanation of the stability in terms of the inter-atomic bonding, which ultimately determine the numerical results. Thus, a bonding model similar to the 2c2e σ bonding model is needed to understand the borane chemistry. With a simple chemical bonding model, one can more easily understand the stability of larger and more complex boron compounds and can further predict and design the boron containing structures for specific applications.

In order to rationalize Wade's rule, the PSEP model was proposed.[14, 15] In this model, each of the n vertex boron atoms in a *closo* borane contributes three AOs for skeletal bonding: one sp_z hybrid AOs radially pointing to the hollow center of the cluster, and a pair of p AOs oriented tangential to the cluster surface. The $2n$ tangential AOs interact to form n bonding and n antibonding MOs, and the n radial AOs hybridize in-phase to generate only one bonding MO and $n - 1$ antibonding MOs. Consequently, there are $n + 1$ skeletal bonds, together with n B-H bonds, giving rise to $2n + 1$ bonds ($4n + 2$ electrons) for a $B_nH_n^{2-}$ cluster. This bonding model has been used to explain Wade's rule and why all the *closo* boranes should be dianions.

The PSEP has been the most influential model for borane chemistry in the last three decades. The subsequently developed tensor spherical harmonics (TSH) model[16, 17, 18] is often considered as the best support to the PSEP model.[4] The TSH model is essentially a MO calculation using guessed basis wavefunctions that are linear combination of the radial and tangential atomic orbitals (AOs) suggested in the PSEP model. The scalar and vector spherical harmonic functions are used as the coefficients in the MO expansion on the AOs. The diagonolization of the Hamiltonian matrix using the guessed basis wavefunctions gives the correct number of bonding skeletal orbitals for boranes. However, it is misleading to believe that the success of getting such very qualitative result has any implication on the physical significance of the guessed basis wavefunctions. Many different basis functions may be chosen to get the qualitative MO description of the boranes.[19] Thus, the TSH model should be considered as one of the many numerical proofs of the Wade's rule in the literature, but does not provide a support to the PSEP model. In fact, the MOs localized at the vertex boron atoms as proposed by the PSEP model[15, 16, 17] are qualitatively different from the 3c2e MOs. The validity of the 3c2e MOs have been confirmed both experimentally[20] and theoretically[21, 22] for boron cages in different chemical environments.

The validity of the PSEP model also relies on the assumption that there is no mixing between radial and tangential AOs such that all the radial AOs combine in-phase to generate only one bonding MO while all the tangential AOs form n localized electron pairs. Based on this assumption, the *closo* boranes should have $n + 1$ skeletal bonds and should be -2 charged. However, it has been found that there is significant mixing between the radial and tangential states.[17, 23] Taking such state-mixing into account in the TSH model improves the numerical results[17] but also destroy the pictorial simplicity of the PSEP model.

The core state (delocalized near the interior of the cage) in the *closo* boranes is essential in the PSEP model as it provides one bonding orbital (in addition to the n skeletal bonds on the cage surface) to trap two extra electrons to form dianion. However, this core state is actually commonly seen in the neutral cage and tube structures, such as neutral C_{60} and carbon nanotubes. This core state is simply a result of hybridization of all in-plane states caused by the curvature and has nothing to do with the dianion nature of the *closo* boranes. Wade's rule has been shown to even hold for elongated *closo* boranes (tubes with two closed ends), such as $B_{17}H_{17}^{2-}$, $B_{22}H_{22}^{2-}$, and $B_{27}H_{27}^{2-}$,[23] where in-phase hybridization of radial AOs assumed in the PSEP model is impossible. For larger *closo* boranes, such as $B_{32}H_{32}^{2-}$, which also obeys Wade's rule,[24] the $sp2$ hybridization is apparently more appropriate than the assumed sp hybridization in the PSEP model.

Based on the above discussions, we find that the PSEP model gives an incorrect chemical bonding picture for boranes, and thus cannot be used to rationalize Wade's rule.
The central question for the bonding in *closo* boranes is how $3n$ AOs interact to form $n + 1$ skeletal bonds that hold $2n + 2$ skeletal electrons.

THEORY

The skeletal bonding in the triangular boron lattice is metallic in nature. The electron distribution on a 2D boron sheet with triangular lattice exhibits a nearly uniform distribution in the interstitial region.[25] The 2D boron sheet and 3D *closo* borane cluster both have the same underlying triangular lattice. However, the boron sheet is neutral whereas the *closo* borane is dianion. Thus, the dianion character of the *closo* boranes may be related to the topological transformation from 2D sheet to 3D cage. For a boron atom on a 2D sheet, one valence electron participates in the π bonding and two other valence electrons participate in skeletal bonding. Consequently, on a 2D boron triangular lattice with n boron atoms and $2n$ triangles, there are $2n$ electrons that participate in the skeletal bonding, leading to an optimal electron density of one electron on every triangle. For a *closo* borane ($B_nH_n^{2-}$), n boron atoms form only $2n-4$ triangles[2] instead of $2n$ triangles in the case of 2D boron sheet. The reduced number of triangles is due to the reduced boron coordination numbers in cage structures compared to that for 2D sheet. In analogy to the 2D boron sheet, the $2n-4$ triangles can hold $2n-4$ skeletal electrons. Based on both experimental and theoretical observations of triangularly shaped wavefunctions in *closo* boranes, we adopt a 3c2e bonding picture for the surface of *closo* boranes, which requires $3n-6$ AOs to form $n-2$ bonding orbitals accommodating $2n-4$ electrons. However, the $n-2$ bonding orbitals covers only half of the $2n-4$ triangles in *closo* boranes. The observed relatively uniform electron distribution on the *closo* borane surface represents a lower-energy electron distribution as a result of resonance between two equivalent electron distributions each of which has half triangles covered by 3c2e bonds, in analogy to the resonance in benzene. Thus, we assign $3n-6$ AOs in $B_nH_n^{2-}$ for $n-2$ resonant 3c2e MOs holding $2n-4$ electrons. Since there are $3n$ AOs and $2n+2$ electrons available for skeletal bonding in a $B_nH_n^{2-}$, there are 6 AOs and 6 electrons that are not accounted for by the resonant 3c2e bonding. These 6 AOs can thus form 3 MOs to accommodate exactly the remaining 6 electrons to form a close-shell electronic structure. Note that these 3 additional MOs cannot be 2c2e bonds. Otherwise, any resulted electron distribution would be inconsistent with the calculated symmetry of the *closo* boranes. As we suggested above, the skeletal bonding is metallic, these three additional MOs should be globally delocalized (GD) metallic states. The space reduction associated with the topological transformation from 2D sheet to 3D cage expels these three MOs out of the resonant 3c2e bonding network on the *closo* borane surface in order to reduce the kinetic energy.

To illustrate the above bonding model, we take the most stable *closo* borane, $B_{12}H_{12}^{2-}$, as an example. According to our model, there should be 12 B-H, 10 resonant 3c2e, and 3 GD orbitals in $B_{12}H_{12}^{2-}$. We can identify these states in the electronic structure calculated based on density functional theory (see Fig. 2). The states shown in solid lines in Fig. 2 are B-H states and those in dotted lines are skeletal states. Among skeletal states, G_u and H_g states can be easily identified as resonant 3c2e states. The A_g skeletal state is delocalized inside the cage as a result of hybridization of all surface resonant 3c2e states, commonly seen in caged structures. The T_{1u} skeletal state is the GD state with no 3c2e or 2c2e character. Since these GD orbitals are deeply bound, they will not be significantly destabilized by size and symmetry reduction in other *closo* boranes and will always trap two extra electrons to form a close shell. This explains why all the *closo* boranes should be dianions. The chemical bonding model in *closo* boranes is summarized in Table I.

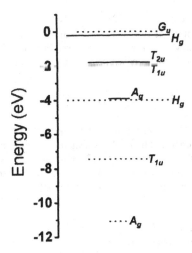

Figure 2. Energy levels of icosahedral $B_{12}H_{12}^{2-}$ with the energy of the highest occupied molecular level set to zero. The energy levels in dotted line indicate the skeletal states, i.e., resonant 3c2e (A_g, H_g, and G_u states) and the globally delocalized state (T_{1u} state), while the levels in solid line indicate usual 2c2e B-H bonding states.

Table I. Number of the atomic (N_{AO}) and molecular (N_{MO}) orbitals for skeletal and B-H bonding in a $B_nH_n^{2-}$ cluster. The skeletal orbitals consist of 3c2e and some additional globally delocalized (GD) orbitals. Number of electrons (N_e) in each category is also shown. The existence of ($n + 1$) skeletal MOs is known as Wade's rule.

	Skeletal			B-H	Total
	Resonant 3c2e	GD states	Subtotal		
N_{AO}	$3n - 6$	6	$3n$	$2n$	$5n$
N_{MO}	$n - 2$	3	$n + 1$	n	$2n + 1$
N_e	$2n - 4$	6	$2n + 2$	$2n$	$4n + 2$

CONCLUSIONS

In summary, we show that there are three types of chemical bonds in *closo* boranes: (1) the usual 2c2e B-H bonding, (2) the delocalized resonant 3c2e skeletal bonding, and (3) the additional globally delocalized states due to space curving and the resulted reduction of boron coordination numbers in a cage from that of 2D triangular lattice. The deeply bound GD states need to trap two extra electrons to be fully occupied, which explains why all the $B_nH_n^{2-}$ are dianions. With this new bonding model for *closo* boranes, the chemical bonding for *closo*, *nido*, and *arachno* boranes can be understood based on a combination of 3c2e (resonant and non-resonant) and 2c2e bonding.

ACKNOWLEDGMENTS

We thank Mark R. Pederson, Zhongfang Chen, and David J. Singh for helpful discussions. This work was supported by the U. S. Department of Energy (DOE), BES and EERE, under Contract No. DE-AC39-98-GO10337. Work at ORNL was supported by DOE Office of Nonproliferation Research and Development NA22. SS acknowledges the financial support from Ministry of Education, Science and Culture of Japan.

[1] G. N. Lewis, J. Am. Chem. Soc. **38**, 762 (1916)

[2] R. B. King, Chem. Rev. **101**, 1119 (2001).

[3] E. D. Jemmis, M. M. Balakrishnarajan, and P. D. Pancharatna, Chem. Rev. **102**, 93 (2002).

[4] M. A. Fox and K. Wade, Pure Appl. Chem. **75**, 1315 (2003).

[5] R. E. Williams, Inorg. Chem. **10**, 210 (1971).

[6] W. H. Eberhardt, B. Crawford, and W. N. Lipscomb, J. Chem. Phys. **22**, 989 (1954).

[7] W. N. Lipscomb, Science **196**, 1047 (1977).

[8] M. E. O'Neill and K Wade, Inorg. Chem. **21**, 461 (1982); M. E. O'Neill and K Wade, Polyhedron **3**, 199 (1984).

[9] K. Wade, Chem. Commun. 792 (1971).

[10] H. C. Longuet-Higgins and M. de V. Roberts, Proc. Roy. Soc. **230A**, 110 (1955).

[11] R. Hoffmann and W. N. Lipscomb, J. Chem. Phys. **36**, 2179 (1962).

[12] W. N. Lipscomb, *Boron Hydride*, W. A. Benjamin: New York, 1963.

[13] R. B. King and D. H. Rouvray, J. Am. Chem. Soc. **99**, 7834 (1977).

[14] K. Wade, Adv. Inorg. Chem. Radiochem. **18**, 1 (1976).

[15] R. D. Gillespie, W. W. Porterfield, and K. Wade, Polyhedron **6**, 2129 (1987)

[16] A. J. Stone, Inorg. Chem. **20**, 563 (1981).

[17] A. J. Stone and M. J. Alderton, Inorg. Chem. **21**, 2297 (1982)

[18] P. W. Fowler and W. W. Porterfield, Inorg. Chem. **24**, 3511 (1985).

[19] S. F. A. Kettle, V. Tomlinson, J. Chem. Soc. A 2002 (1969); S. F. A. Kettle, V. Tomlinson, J. Chem. Soc. A 2007 (1969)

[20] M. Fujimori, T. Nakata, T. Nakayama, E. Nishibori, K. Kimura, M. Takata, and M. Sakata, Phys. Rev. Lett. **82**, 4452 (1999).

[21] R. B. King, T. Heine, C. Corminboeuf, and P. v. R. Schleyer, J. Am. Chem. Soc. **126**, 430 (2004).

[22] J. Yamauchi, N. Aoki, and I. Mizushima, Phys. Rev. B **55**, R10245 (1997).

[23] M. M. Balakrishnarajan, R. Hoffmann, P. D. Pancharatna, and E. D. Jemmis, Inorg. Chem. **42**, 4650 (2003).

[24] A. A. Quong, M. R. Pederson, and J. Q. Broughton, Phys. Rev. B **50**, 4787 (1994).

[25] M. H. Evans, J. D. Joannopoulos, and S. T. Pantelides, Phys. Rev. B **72**, 045434 (2005).

Mater. Res. Soc. Symp. Proc. Vol. 1038 © 2008 Materials Research Society   1038-O05-09

Computational Models for Crystal Growth of Radiation Detector Materials: Growth of CZT by the EDG Method

Jeffrey J. Derby, and David Gasperino
Chemical Engineering and Materials Science, University of Minnesota, 421 Washington Ave,
SE, Minneapolis, MN, 55455-0132

ABSTRACT

Crystals are the central materials element of most gamma radiation detection systems, yet there remains surprisingly little fundamental understanding about how these crystals grow, how growth conditions affect crystal properties, and, ultimately, how detector performance is affected. Without this understanding, the prospect for significant materials improvement, i.e., growing larger crystals with superior quality and at a lower cost, remains a difficult and expensive exercise involving exhaustive trial-and-error experimentation in the laboratory. Thus, the overall goal of this research is to develop and apply computational modeling to better understand the processes used to grow bulk crystals employed in radiation detectors. Specifically, the work discussed here aims at understanding the growth of cadmium zinc telluride (CZT), a material of long interest to the detector community. We consider the growth of CZT via gradient freeze processes in electrodynamic multizone furnaces and show how crucible mounting and design are predicted to affect conditions for crystal growth.

INTRODUCTION

Large, single crystals of cadmium zinc telluride (CZT) form the heart of several advanced gamma detectors, which promise portable, low-cost, and sensitive devices to monitor radioactive materials [1-6]. Decades of development have produced great strides in improved crystal growth processes and better materials for these systems [7,8], and CZT crystals of sufficient quality are now commercially available for simple counting and monitoring applications. However, today's homeland security needs demand large field-of-view imaging and high-sensitivity, high-resolution spectroscopic analysis, which require large, single CZT crystals with spatially uniform charge-transport properties [8]. Affordable material of this size and quality is not yet available.

The growth of large CZT crystals is not well understood and surprisingly less mature than semiconductor crystal growth employed for the electronics industry. One reason for this state of affairs is that CZT crystal growth is far more challenging than that of more traditional semiconductor crystals, such as silicon and gallium arsenide. Indeed, the growth of large, single crystals of CdTe or CZT is notoriously difficult; Rudolph [9,10] details the many challenges encountered during growth. The end result for the growth of CZT is that typical yields of useable material from a crystalline boule remain at 10% or lower [11], resulting in very high materials costs. To improve existing radiation detector crystals or to develop new materials, the performance-property-processing loop must be closed. Namely, device performance must be understood in terms of a mechanistic understanding of crystal growth, and this understanding must be put into the practice of crystal growth using new approaches and modern ideas.

Toward the goal of obtaining a more complete understanding of crystal growth, we are developing and applying computational models of crystal growth processes. Relatively few prior modeling studies have addressed the vertical Bridgman growth of CdTe and CZT. Sen *et al.* [12], Pfeiffer and Mühlberg [13], and Parfeniuk *et al.* [14], employed models that neglected melt convection; however, because of the relatively large influence of convection on heat transfer in this system, this is a poor assumption. We have developed more detailed two-dimensional models to study the application of both vertical [15-18] and horizontal [19-21] Bridgman methods for the production of CZT infrared detector substrates. We followed this work with analyses of high-pressure growth system for CZT radiation detector crystals [22-24]. Yeckel *et al.* [25] studied the three-dimensional effects on melt convection and solute segregation caused by system imperfections in CZT crystal growth in a vertical Bridgman system but employed an idealized representation of furnace heat transfer. Recently, we have used a coupled modeling approach to analyze an industrial system, used by eV Products, Inc. This modeling approach employs a global-scale model for furnace heat transfer coupled to a local-scale model for heat transfer, melt flow, and solidification within the ampoule [26]. Such an approach is needed to relate model predictions to actual process operation. With this approach, we have reported on a successful validation of our model with experimental measurements [27]. In more recent work, we have investigated the feasibility of using furnace temperature profile manipulation in an EDG furnace to control interface shape during growth of CZT [28,29].

METHODS AND APPROACH

For the results presented here, we employ the crystal growth simulation software CrysMAS, developed by the Crystal Growth Laboratory of the Fraunhofer Institute of Integrated Systems and Device Technology (IISB) in Erlangen, Germany [30,31]. This package is capable of predicting high temperature heat transfer within complex crystal growth furnaces by solving the energy conservation equations using the finite volume method on an unstructured triangular grid. The radiative heat transfer calculation is implemented using view factors with an enclosure method. CrysMAS applies a structured grid to perform the heat transfer, fluid flow, and phase change computations needed within the ampoule. Furnace setpoint temperatures are specified, and the heater powers are solved as unknowns. CrysMAS employs a quasi-Newton iterative method to arrive at a converged solution.

Our model is based on a Mellen Company electrodynamic gradient freeze (EDG) furnace with 18 controlled heating zones. The axial symmetry of the experimental furnace allows for a simplified two-dimensional model representation in cylindrical coordinate space. Computations are performed specifically for the design and operation of furnaces employed by the groups of Professor Kelvin Lynn of Washington State University (WSU) and Dr. Mary Bliss of Pacific Northwest National Laboratories (PNNL). We report on model development, validation, and more results in [32,33]. A schematic diagram of the computational domain for the model and a representation of the finite volume mesh are shown Fig. 1.

Several meshes were constructed and used to assess numerical convergence and solution accuracy. Accurate computations were achieved with approximately 115,000 degrees of freedom for the PBN ampoule system and 170,000 degrees of freedom for the graphite ampoule system. Solution times for these systems varied, respectively, from approximately 45 seconds per iteration to approximately 140 seconds per iteration on a Dell Precision Workstation outfitted with two Quad Core Intel Xeon Processors running at 2.66GHz.

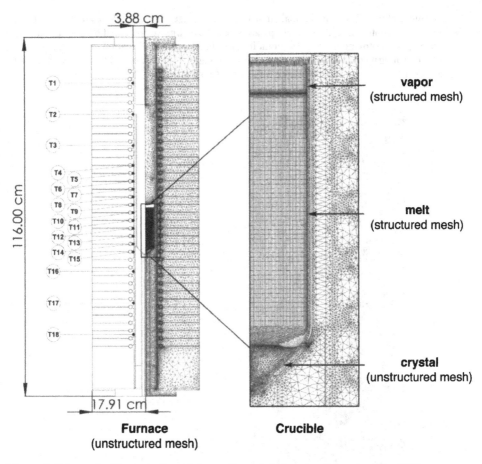

3.88 cm

T1
T2
T3
T4 T5
T6 T7
T8 T9
T10 T11
T12 T13
T14 T15
T16
T17
T18

116.00 cm

17.91 cm

Furnace
(unstructured mesh)

vapor
(structured mesh)

melt
(structured mesh)

crystal
(unstructured mesh)

Crucible

Figure 1: Schematic diagram of crystal growth model (showing the case of the PBN ampoule) constructed using CrysMAS.

RESULTS

The global temperature field is shown in Figure 2 along with a schematic representation of how the gradient freeze method is employed. In the bore of the furnace, an ampoule is placed that contains the charge of cadmium, zinc, and tellurium in proper ratios. After bringing the entire contents of the ampoule to above the melting point of CZT (1365 K), a vertical thermal gradient is established over the length of the ampoule, with lower temperatures below and higher temperatures above. Then the thermal environment is changed in a time-dependent manner to effectively translate the melting point upward along the charge, as indicated by the right-hand side of Figure 2. Moving the melting point induces directional solidification of the charge and enables crystal growth. The EDG system employs computer control of individual heating zones to impose a specific temperature setpoint schedule for growth.

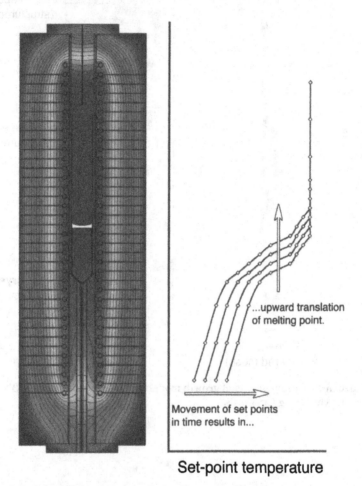

Set-point temperature

Figure 2: The global temperature field of the electrodynamic gradient freeze furnace is dynamically adjusted using heat input to individual zone to induce directional solidification of the charge in the ampoule.

In the computations presented here, we consider the outcome from two design changes to the growth system, as depicted in Figure 3. The first change (shown on the left side of the figure) involves a modification of the ampoule support system meant to increase the axial flow of heat through the bottom center of the ampoule. The silicon carbide (SiC) support rod is extended completely through the refractory material plugging the bottom of the furnace bore. This change allows for a heat path through the relatively high thermal conductivity SiC rod out of the high-temperature region of the furnace to the cooler ambient. The other modification adds a graphite disk under the tip of the ampoule to maintain intimate thermal contact between the ampoule and the support system. The next design change is more substantial and indicated on the right of Figure 3. Here, we wish to predict the impact on growth conditions by replacing the existing graphite ampoule with one fabricated from pyrolitic boron nitride (PBN).

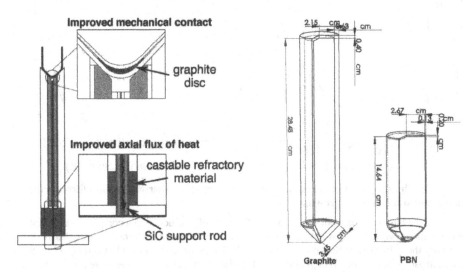

Figure 3: Two design modifications to the crystal growth system are considered in the computations presented here. The first, depicted on the left, is a change to the ampoule support system meant to increase the axial flow of heat. The second, shown on the right, is a change to the shape and composition of the ampoule.

Change of ampoule support

Profiles of the temperature field along the system centerline, through the molten charge, are shown in Figure 4 for both ampoule types for the original furnace design (old) and the new ampoule support configuration (as indicated on the left-hand side of Figure 3). Note that the temperature field is plotted on the abscissa while vertical position (normalized so that the origin corresponds to the bottom, interior tip of the ampoule) is plotted on the ordinate. The temperature curves correspond to furnace conditions (by vertical translation of the set points) that place the melting point of CZT at the bottom tip of the original graphite ampoule. These furnace set points are then employed for all of the other design configurations shown in Figure 4.

It is instructive to compare these profiles case by case. For the graphite ampoule, shown in the plot on the left of Figure 4, there is little change in the centerline temperature profile due to the design changes. In fact, the modifications result in an effect opposite to that desired. The new design results in higher temperatures in the tip region of the ampoule and shallower axial thermal gradients. A more detailed analysis of heat transfer shows that the graphite disk near the ampoule tip increases the radial conduction of heat to the ampoule; this increase of heat flowing radially inward to the charge overwhelms any increases in the axial heat flux out of the charge through the tip. This mechanism is discussed more completely in [33].

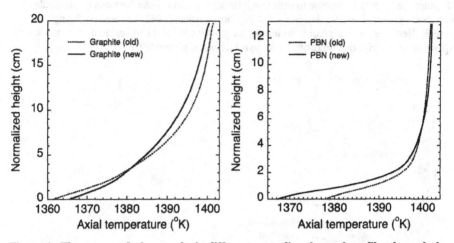

Figure 4: The support design results in different centerline thermal profiles through the melt prior to the onset of growth.

The ampoule support design changes affect the PBN crucible in a completely different manner, as shown in the right-side plot of Figure 4. In this system, the new support clearly increases the axial flow of heat out of the ampoule tip and produces much lower temperatures and increased axial temperature gradients. (Note that a steeper axial temperature profile appears flatter in the plots shown here, with temperature on the abscissa and position on the ordinate.) This is a generally desirable outcome which will be discussed further in the conclusions section.

Change of ampoule material

A more significant effect in the crystal growth system is achieved by changing the ampoule design (as indicated by the right pane of Figure 3). Figure 5 depicts both systems, with temperature contours shown on the left-side and streamfunction contours for flow within the melt shown on the right. The velocity of the melt is everywhere tangent to the streamfunction contours (also known as streamlines), and the volumetric flow is proportional to the differences between the streamfunction values. For equally spaced streamfunction contours, as are shown here, the flow is stronger where the streamlines are more closely spaced. All other features of the system are unchanged for these two computations, and we consider the design of the new support system indicated by the left pane of Figure 3. In particular, both of these systems employ the same furnace set-point profiles. Each state corresponds with a shifted furnace setpoint profile that yields the freezing point of CZT at the bottom interior of the ampoule. In these unseeded systems, this point corresponds to just prior to the nucleation of solid material and growth.

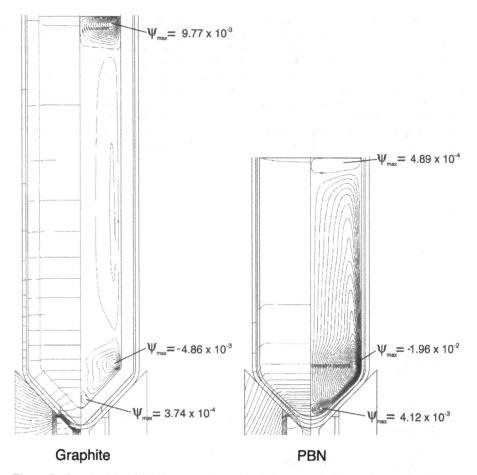

$\Psi_{max}= 9.77 \times 10^{-3}$

$\Psi_{max}= 4.89 \times 10^{-4}$

$\Psi_{max}=-4.86 \times 10^{-3}$

$\Psi_{max}= -1.96 \times 10^{-2}$

$\Psi_{max}= 3.74 \times 10^{-4}$

$\Psi_{max}= 4.12 \times 10^{-3}$

Graphite   PBN

Figure 5: Comparison of the state of each system prior to nucleation and growth. All system parameters are held constant except for the crucible design. Temperature contours are plotted with a spacing of 2 K on the left of each image, and streamfunction contours are spaced at 7 cm^2/s on the right.

For the system with the graphite ampoule, the temperature profile through the melt is nearly linear in the axial direction, as indicted by the nearly constant vertical spacing of the temperature contours; however, the gradient increases slightly as the cone region is approached. In the melt, a complicated, multiple-cell flow structure is driven by buoyancy. However, in the tip region and through most of the bulk of the melt, the flow is relatively weak. This flow is strong enough for convective effects to flatten the temperature contours in the radial direction. Near the top of the melt, temperature-dependent surface tension drives relatively intense, Marangoni flows; however, this flow cell is placed far enough away from the cone region to be unimportant for the initial stages of growth.

The situation is markedly changed for the system with the PBN ampoule, shown on the right of Figure 5. First, there is a significant increase in the thermal gradient though the cone region of the ampoule, while much of the remaining melt is nearly isothermal. The difference in the centerline temperatures is shown in Figure 6 for the two cases. (Again, a flatter curve on this plot corresponds to a steeper axial gradient.) Observing the temperature contours, one also sees the influence of the strong melt flow, which causes a boundary layer to form near the ampoule walls and deflects the isothermal temperature contours there. The buoyant flow in the melt is much more intense, nearly an order of magnitude faster and stronger than that in the melt of the graphite ampoule. In contrast, the Marangoni flow driven by surface tension gradients along the free surface at the top of the melt is much weaker than in the graphite case.

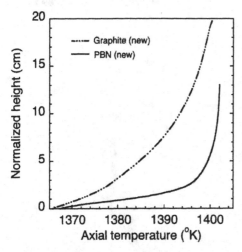

Figure 6: **Axial temperatures along the centerline are very different for the graphite and PBN crucibles.**

DISCUSSION AND CONCLUSIONS

We have performed detailed computations of the electrodynamic gradient freeze CZT growth systems employed by Bliss (PNNL) and Lynn (WSU) using CrysMAS. After validation [32,33], we have applied the model to predict system behavior after design changes for the ampoule support system and the ampoule itself.

The first computations assessed the effects of system ampoule support changes meant to increase the axial heat flux through the charge. In the heat transfer calculations for the graphite ampoule system, we found a surprising, counterintuitive result; the new support design did not increase the axial gradient in this growth system. Rather, the axial temperature profile became shallower due to a subtle change in the balance between radial and axial heat fluxes. In contrast, the same design changes did increase axial thermal gradients in the PBN system. The engineering of a mostly axially aligned flow of heat is generally preferred in vertical melt growth systems, because axial gradients are responsible for driving the directional solidification of crystalline material. Radial gradients more strongly affect the shape of the growth interface and thermoelastic stresses experienced by the crystal during growth. Considering these criteria, the PBN ampoule with the new ampoule support design produces thermal conditions that are better suited to growth than the other cases.

The second set of computations considered how ampoule material and design affect the thermal environment in the growth system. Significant changes in thermal conditions are predicted for the two ampoules, especially for the axial thermal profiles through the melt. While axial gradients were nearly constant for the graphite ampoule case, the thermal field through the melt contained in the PBN ampoule was nearly isothermal though much of the melt but steeply changing axially through the cone region of the ampoule. A possible outcome of this thermal field in the PBN system is the establishment of more favorable conditions for controlling the nucleation of solid. While we did not attempt to simulate nucleation, it is apparent that the PBN ampoule produces a thermal field through the cone region of the ampoule that has a significantly larger gradient. Thus, undercooled melt, which is the prelude to nucleation, is constrained to a much smaller volume in the PBN system than in the graphite system. This is expected to be important to minimize the likelihood of multiple nucleation events, which would lead to multiple grains during subsequent solidification.

A more detailed analysis of the different systems, as is carried out in [32,33], shows that the thick, highly conductive graphite ampoule wall channels a significant amount of heat through the furnace, making geometrical changes, like the cone region of the ampoule, relatively unimportant in affecting overall heat flow and the temperature field though the melt. On the other hand, the much lower thermal conductivity of the PBN ampoule and its thin walls substantially reduce the heat flow carried by the ampoule. In the PBN system, the cone region geometry acts to focus axial heat flow out of the ampoule tip.

We believe that progress in crystal growth will be accelerated by the careful application of computational models. Clarifying the conditions of crystal growth and assessing the impact of process changes will promote rational decisions for process improvement. Indeed, computations such as performed here are able to predict the impact of design changes far more cheaply and quickly than experiments.

However, much work remains. The fundamental links between macroscopic crystal growth conditions and crystal structure and properties are still poorly understood, especially with regard to microstructural properties of CZT, such as second-phase (tellurium) particle formation. More fundamental studies of the materials science of CZT crystal growth are needed. In addition, the development of advanced, real-time, model-based control methods will be needed to improve the quality and reduce the costs associated with detector-grade CZT production. The development of realistic, physically faithful models is a first step toward both of these objectives.

ACKNOWLEDGMENTS

This material is based upon work supported in part by the Department of Energy, National Nuclear Security Administration, under Award Number DE-FG52-06NA27498, the content of which does not necessarily reflect the position or policy of the United States Government, and no official endorsement should be inferred. Computational resources were provided by the Minnesota Supercomputing Institute. The authors would like to acknowledge the significant input to this work of Mary Bliss (Pacific Northwest National Laboratory), Kelly Johnson and Kelvin Lynn (Washington State University) and ongoing collaboration on crystal growth modeling software with Jochen Friedrich, Georg Müller, and Thomas Jung (Crystal Growth Laboratory, Fraunhofer Institute IISB).

REFERENCES

1. 1. E. Raiskin and J.F. Butler. *IEEE Trans. Nuclear Science*, 35:81, 1988.
2. J.F. Butler, C.L. Lingren, and F.P. Doty. *IEEE Trans. Nucl. Phys.*, 39:605, 1992.
3. F.P. Doty, J.F. Butler, J.F. Schetzina, and K.A. Bowers. *J. Vac. Sci. Technol. B*, 10:1418, 1992.
4. J.F. Butler, B. Apotovsky, A. Niemela, and H. Sipila. In *Proceedings of the SPIE*, volume 2009, page p. 121. SPIE, Bellingham, WA, 1993.
5. R.B. James, T.E. Schlesinger, J. Lund, and M. Schieber. In T.E. Schlesinger and R.B. James, editors, *Semiconductors for Room Temperature Nuclear Detector Applications*, volume 43, page p. 335. Academic Press, San Diego, 1995.
6. R.B. James and P. Siffert, editors, "Room Temperature Semiconductor Detectors: Proceedings of the 11th International Workshop on Room Temperature Semiconductor X- and Gamma-Ray Detectors and Associated Electronics,"*Nuclear Instruments and Methods in Physics Research A*, volume 458, 2001.
7. Szeles, C. Cameron, S.E. Ndap, J.-O. Chalmers, W.C. *IEEE Trans. Nuclear Science*, 49:2535, 2002.
8. C. Szeles, S.E. Cameron, S. A. Soldner, J.-O. Ndap, And M. D. Reed, *Journal of ELECTRONIC MATERIALS* Vol.33, 742–752 (2004).
9. P. Rudolph, Progr. Crystal Growth and Charact. 29, 275 (1994).
10. P. Rudolph. In M. Isshiki, editor, *Recent Development of Bulk Crystal Growth*. Research Signpost, Trivandrum, India, 1998.
11. J.J. Griesmer, B. Kline, J. Grosholz, K. Parnham, D. Gagnon, In: Proceedings of IEEE MIC 2001, San Diego, Nov. 2001.
12. S. Sen, W.H. Konkel, S.J. Tighe, L.G. Bland, S.R. Sharma, and R.E. Taylor. *J. Crystal Growth*, 86:111-117, 1988.
13. M. Pfeiffer and M. Mühlberg. *J. Crystal Growth*, 118:269, 1992.
14. C. Parfeniuk, F. Weinberg, I.V. Samarasekera, C. Schvezov, and L. Li. *J. Crystal Growth*, 119:261, 1992.
15. S. Kuppurao, S. Brandon, and J.J. Derby, *J. Crystal Growth* 155, 93–102 (1995).
16. S. Kuppurao, S. Brandon, and J.J. Derby, *J. Crystal Growth* 155, 103–111 (1995).
17. S. Kuppurao, S. Brandon, and J.J. Derby, *J. Crystal Growth* 158, 459–470 (1996).
18. S. Kuppurao and J.J. Derby, *J. Crystal Growth* 172, 350–360 (1997).
19. K. Edwards and J.J. Derby. *J. Crystal Growth*, 179, 120, 1997.
20. K. Edwards and J.J. Derby. *J. Crystal Growth*, 179, 133, 1997.
21. K. Edwards and J.J. Derby. *J. Crystal Growth*, 206, 37–50, 1999.
22. A. Yeckel, F.P. Doty, and J.J. Derby, J. Crystal Growth 203, 87–102 (1999).
23. A. Yeckel and J.J. Derby, J. Crystal Growth 209, 734–750 (2000).

24. A. Yeckel and J.J. Derby, J. Crystal Growth 233, 599–608 (2001).

25. A. Yeckel, G. Compere, A. Pandy, and J.J. Derby. *J. Crystal Growth*, 263:629–644, 2004.

26. A. Yeckel, A. Pandy, and J.J. Derby, *Int. J. Numer. Meth. Engng.* 67, 1768–1789 (2006).

27. Pandy, A. Yeckel, M. Reed, C. Szeles, M. Hainke, G. Müller, and J.J. Derby, *J. Crystal Growth* 276, 133–147 (2005).

28. L. Lun, A. Yeckel, C. Szeles, M. Reed, P. Daoutidis, and J.J. Derby, *J. Crystal Growth* 290, 35–43 (2006).

29. L. Lun, A. Yeckel, J.J. Derby, and P. Daoutidis, in: *Proceedings of the IEEE 2007 Mediterranean Conference on Control and Automation (MED 2007)*, Athens, Greece, June 27–29, 2007.

30. M. Kurz, A. Pusztai, and G. Müller. *J. Crystal Growth*, 198:101, 1999.

31. R. Backofen, M. Kurz, and G. Müller. *J. Crystal Growth*, 199:210, 2000.

32. D. Gasperino, K. Jones, K. Lynn, M. Bliss, and J.J. Derby, *J. Crystal Growth*, to be submitted, 2008.

33. D. Gasperino, Ph.D. thesis, University of Minnesota, in preparation.

Mater. Res. Soc. Symp. Proc. Vol. 1038 © 2008 Materials Research Society

Recent Advances in Ceramic Scintillators

Edgar V Van Loef[1], Yimin Wang[1], Jarek Glodo[1], Charles Brecher[2], Alex Lempicki[2], and Kanai S Shah[1]

[1]RMD, 44 Hunt Street, Watertown, MA, 02472
[2]ALEM Associates, 44 Hunt Street, Watertown, MA, 02472

ABSTRACT

A review is presented of recent ceramic scintillator R&D. Attention is focussed on Ce doped gamma-ray scintillators for medical imaging applications. Ceramic scintillators discussed in detail include $SrHfO_3$:Ce and $Lu_2Hf_2O_7$:Ce. These materials combine a high density and high atomic number with fast emission and a good light yield and may find practical application in medical imaging modalities such as Positron Emission Tomography and Computed Tomography.

INTRODUCTION

Inorganic scintillators coupled to optical detectors such as photomultiplier tubes (PMTs) or silicon photodiodes provide detection and spectroscopy of ionizing radiation and charged particles by converting ionizing radiation into optical photons, which are subsequently detected by the PMT or photodiode. These devices are an important part of medical imaging applications such as positron emission tomography (PET) and computed tomography (CT), find practical use in nuclear and particle physics experiments, and are indispensable for nuclear non-proliferation.

While most efforts have been directed towards the research and development of novel inorganic single crystals for scintillation detection, the field of ceramic scintillators has only received scant attention since the development of the HiLight™ scintillator by General Electric. This is in part due to the difficult task of fabricating transparent ceramics from non-cubic materials and the manufacturing of large volume devices. Recently however, ceramic scintillators are gaining interest because they may be produced in transparent forms from non-cubic materials using nanotechnology and alternative chemical synthesis routes.

In this paper we will present a historical overview of ceramic scintillators. Scintillation properties of various translucent/transparent ceramic scintillators are reported. We will limit the discussion of novel transparent ceramics to the manufacturing and characterization of $SrHfO_3$:Ce and $Lu_2Hf_2O_7$:Ce ceramics. Conventional and novel chemical methods to synthesize powders are described and techniques to manufacture ceramic scintillators such as hot pressing and hot isostatic pressing are detailed. Finally, the optical and scintillation properties by means of radioluminescence, pulse height, luminescence decay and timing measurements are presented.

HISTORY

The use of polycrystalline transparent ceramics for scintillation detection is a rather novel approach, no older than a few decades. A graphical overview of the history of ceramic scintillators is shown in figure 1. Whereas the first inorganic single crystal scintillators were explored during the early 40's and 50's in conjunction with the development of the photomultiplier tube [1], the first polycrystalline ceramic scintillators were developed at a much later stage using technology on transparent ceramics employed previously for lighting applications such as the Lucalox™ high pressure sodium lamp [2,3]. The commercial introduction of the HiLight™ scintillator by General Electric in the early 80' [4,5] triggered a burst of exploration that yielded a few commercially viable ceramic scintillators, such as Gd_2O_2S:Pr,Ce,F and Gd_2O_2S:Pr (UFC), developed by Hitachi Metals [6-8] and Siemens [9,10], respectively.

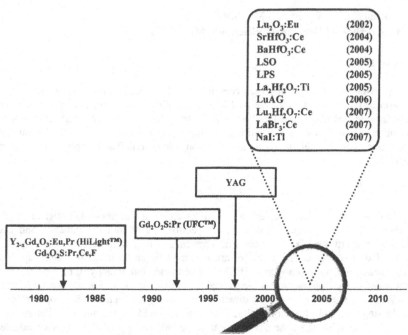

Figure 1. History of ceramic scintillators

However, most of the ceramic scintillators discovered during the late 80's and early 90's were either translucent or opaque. Because of this, ceramic scintillators did not find widespread application in medical imaging devices and were exclusively used for X-ray Computed Tomography (CT). Nevertheless, the advent of transparent YAG [11] prompted a renaissance in the research and development of optically transparent ceramic scintillators. In the last five years alone, numerous ceramic scintillators have been described that have a high density, fast emission and good light yield, in combination with reasonable optical clarity. Whereas transparency of ceramic materials was usually restricted to cubic materials, nanotechnology and alternative chemical synthesis routes made the production of transparent ceramic scintillators from non-cubic materials feasible.

Many of the new ceramic scintillators are high density oxides that have high melting points prohibiting cost-effective crystal growth. However, the consolidation of powder into a fully dense ceramic provides a low cost and reliable alternative to the growth of single crystals. If ceramic scintillators can be produced in large quantities at low cost and with good optical clarity, ceramic scintillators should become promising candidates for numerous radiation detection applications, including medical imaging modalities such as PET and CT.

NEW CERAMIC SCINTILLATORS

In Table I we compiled information on several new ceramic scintillators that have appeared in literature during the last five years. This list is by no means complete, but paints a picture of what is a happening in ceramic scintillator research. Clearly, most efforts have been directed towards the consolidation of high density oxides into transparent ceramics. One of the first ceramic scintillators in this respect is Lu_2O_3:Eu [12], which can be made into fully transparent specimens. Lutetium oxide doped with europium is a cubic rare-earth oxide with a density of 9.4 g/cm^3, a light output which is more than twice than that of BGO, and has a narrow band emission at 611 nm. Unfortunately, the emissions decays relatively slowly (~ 1 ms) and therefore disqualifies Lu_2O_3:Eu for high-speed medical applications.

Ceramic scintillators made by hot pressing from Lu_2SiO_5 (LSO) and $Lu_2Si_2O_7$ (LPS) powders show favorable scintillation properties compared to the single crystalline form, but due to the monoclinic crystal structure LSO and LPS cannot be formed into fully transparent ceramics by conventional means [13,14]. The combined effects of their optical anisotropy and a grain size on the order of 10 μm produce translucent rather than fully transparent ceramic scintillators. Nevertheless, with appropriate control of the microstructure, light scattering from optical anisotropy can be reduced so much as to allow even such non-cubic materials to be made optically transparent. This is work in progress.

Recently, novel techniques such as a low-temperature combustion process have been successfully employed to produce precursor powders for the consolidation into transparent ceramics. For example, a specimen of $La_2Hf_2O_7$ [15] doped with Tb [16] or Ti [17] shows about 60% transmittance partly due to the highly sinterable, ultra-fine combustion-made precursor powders. Other transparent ceramic scintillators manufactured from powders obtained with this technique include $Lu_3Al_5O_{12}$:Ce [18] and $BaHfO_3$:Ce [19].

Table I. New ceramic scintillators.

Host	Dopant	Crystal system [a]	Density (g/cm^3)	Transparency [b]	Light yield (ph/MeV)	λ_{em} (nm)	τ (ns)
Lu_2O_3	Eu	C	9.4	++	27,000	610	~10^6
Lu_2SiO_5	Ce	M	7.4	+	16,000	420	40
$Lu_2Si_2O_7$	Ce	M	6.2	+	16,000	390	40
$SrHfO_3$	Ce	O	7.6	+	20,000	410	20
$BaHfO_3$	Ce	C	8.4	+	44,000	400	20
$La_2Hf_2O_7$	Ti	C	7.9	++	13,000	475	~10^4
$Lu_2Hf_2O_7$	Ce	C	9.4	++	1,000	400	20
$Lu_3Al_5O_{12}$	Ce	C	6.7	++	3,000	300	100
$LaBr_3$	Ce	H	5.1	+	45,000	385	30
NaI	Tl	C	3.7	+	38,000	415	250

[a] Crystal system: C = cubic, O = Orthorhombic, M = Monoclinic, H = Hexagonal.
[b] Very good/Transparent (++); Reasonable/Translucent (+); Opaque (-).

In the following sections we will discuss in detail the manufacturing and characterization of $SrHfO_3$:Ce and $Lu_2Hf_2O_7$:Ce. These materials combine a high density and high atomic number with fast emission and a good light yield and are the most promising candidates for numerous radiation detection applications, including medical imaging modalities such as PET and CT.

Figure 2. $SrHfO_3$:Ce ceramic.

$SrHfO_3$:Ce

Strontium hafnate belongs to the perovskite type family and has the orthorhombic (Pnma) crystal structure [20]. The density of $SrHfO_3$ is 7.6 g/cm^3 [21]. It has a melting point of about 2730°C [22] and melts congruently. Because of the high temperature, it is very difficult to grow large single crystals. However, $SrHfO_3$ has nearly isotropic optical properties, which allows for the fabrication of fully transparent optical ceramics.

Strontium hafnate precursor powders for ceramic manufacturing are typically synthesized by the solid state technique from $SrCO_3$ and HfO_2. A small amount of $Ce(NO_3)_3 \cdot 6H_2O$ or $Ce_2(CO_3)_3$ is added to provide Ce^{3+} as optically active ion. Since luminescence is only present when Ce is in the trivalent state and located at the Sr site, charge compensation is needed [23]. Therefore, a small amount of $Al(NO_3)_3 \cdot 9H_2O$ or $Al_2(CO_3)_3$ is usually added. Ceramics can be manufactured using a vacuum hot press or a hot isostatic press (HIP) at temperatures between 1400 and 1700°C and pressures up to 30,000 psi [24]. As an example, figure 2 shows a translucent ceramic obtained by hot pressing of $SrHfO_3$:Ce powder. The sample is 1 mm thick and 10 mm in diameter. The color of the ceramics is due to loss of oxygen and the introduction of F-centers. Heat treatment in air restores the oxygen and bleaches the ceramic.

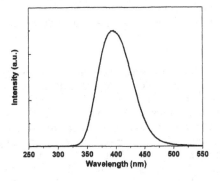

Figure 3. Optical emission spectrum of SrHfO₃:Ce under X-ray excitation.

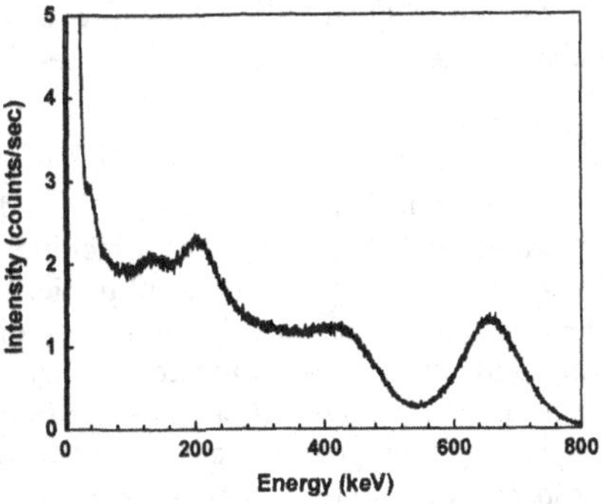

Figure 4. Pulse height spectrum of SrHfO₃:Ce under ^{137}Cs gamma-ray excitation.

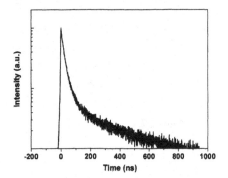

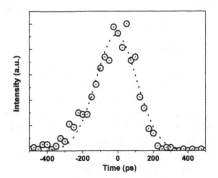

Figure 5. Scintillation decay time spectrum of SrHfO₃:Ce under pulsed X-ray excitation.

Figure 6. Coincidence timing resolution spectrum of SrHfO₃:Ce.

The optical emission spectrum under X-ray excitation is shown in figure 3. The spectrum consists of a broad emission band due to Ce^{3+} emission that peaks at 410 nm. If we compare the integral of the radioluminescence spectra with that of a BGO crystal, the light yield of SrHfO₃:Ce approximately is 20,000 ph/MeV. Figure 4 shows the pulse height spectrum of SrHfO₃:Ce under ^{137}Cs gamma-ray excitation. The light yield is approximately 5000 ph/MeV using a shaping time of 4 µs. The energy resolution R (full-width at half-maximum over peak position) for the 662 keV full energy peak is about 17%. The light yield under gamma-ray excitation is less than that obtained from radioluminescence measurements. The difference may be caused by the presence of long decay components or even afterglow. The scintillation decay time spectrum of SrHfO$_3$:Ce under pulsed X-ray excitation is shown in figure 5. The principal decay time constant is about 20 ns. Figure 6 shows the coincidence timing resolution spectrum of SrHfO$_3$:Ce using Hamamatsu R-5320 PMTs and a ^{68}Ga source. The width of the distribution is 276 ps (full-width at half-maximum over peak position) using a CeBr$_3$ reference crystal. This timing resolution is comparable to the best that can be obtained with high-quality LSO crystals.

Lu$_2$Hf$_2$O$_7$:Ce

Lutetium hafnate crystallizes in a disorder fluorite structure [25] and has a density of approximately 9.4 g/cm^3. It has a melting point of about 2900°C [26] and melts congruently. Precursor powders with very small particle sizes can be obtained by a combustion-type process where lutetium and hafnyl nitrate are dissolved in water and gelated with glycine. The combustion of the power proceeds at about 400°C under the evolution of carbon dioxide, water vapor and nitrous gases. Ceramics can be manufactured by a Sinter-HIP process or by simply hot-pressing the powders at temperatures between 1200 and 1900°C. An example of a transparent Lu$_2$Hf$_2$O$_7$:Ce ceramic is shown in figure 7. The sample is 1 mm thick and 10 mm in diameter. A typical emission spectrum of Lu$_2$Hf$_2$O$_7$:Ce under X-ray excitation is shown in figure 8. The spectrum consists of a broad emission band due to Ce^{3+} emission, peaking at 400 nm. In contrast to SrHfO$_3$:Ce, the light yield of Lu$_2$Hf$_2$O$_7$:Ce is low, about 1000 ph/MeV. The scintillation decay time spectrum of Lu$_2$Hf$_2$O$_7$:Ce under pulsed X-ray excitation is shown in figure 9. The principal decay time constant is about 20 ns.

SUMMARY

In this paper we have presented a historical overview of ceramic scintillators. Scintillation properties of various translucent/transparent ceramic scintillators were reported. Conventional and novel chemical methods to synthesize powders were described and techniques to manufacture ceramic scintillators such as hot pressing and hot isostatic pressing were mentioned. The optical and scintillation properties of SrHfO$_3$:Ce and Lu$_2$Hf$_2$O$_7$:Ce were presented. Several new transparent optical ceramics (TOCs), including SrHfO$_3$:Ce and Lu$_2$Hf$_2$O$_7$:Ce. Many of these TOC scintillators have excellent gamma ray attenuation due to their high density and high effective atomic number, exhibit a fast principle decay component and show light yields up to 40,000 ph/MeV under X-ray excitation. In the case of

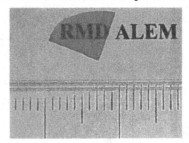

Figure 7. Lu$_2$Hf$_2$O$_7$:Ce ceramic

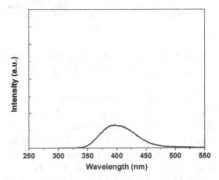

Figure 8. Optical emission spectrum of Lu$_2$Hf$_2$O$_7$:Ce under X-ray excitation.

$Lu_2Hf_2O_7$:Ce, the light yield was rather low. Studies on the fabrication of TOC show that translucent scintillators can be obtained from precursor powders obtained by solid state as well as novel chemical methods such as combustion techniques. If large TOC can be fabricated with improved transparency, reduced afterglow and at low cost, $SrHfO_3$:Ce, $BaHfO_3$:Ce and $Lu_3Al_5O_{12}$:Ce may become of interest for medical imaging applications such as Positron Emission Tomography and Computed Tomography. Additionally, if the light yield of $Lu_2Hf_2O_7$:Ce can be improved, $Lu_2Hf_2O_7$:Ce may also become of interest for selected medical or high energy physics applications.

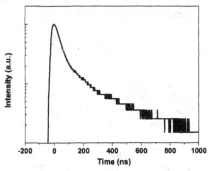

Figure 9. Scintillation decay time spectrum of $Lu_2Hf_2O_7$:Ce under pulsed X-ray

REFERENCES

1. M. J. Weber, "Inorganic scintillators: today and tomorrow," *J. Lumin.*, **100**, 35 (2002).
2. K. Schmidt, "Metal vapor lamps," US Patent, no. 2,971,110 (1961).
3. W.C. Louden, Kurt Schmidt, "High-Pressure Sodium Discharge Arc Lamps," *Illuminating Engineering*, **60**, 696 (1965).
4. Cusano *et al.*, "Rare-earth-doped yttria-gadolinia ceramic scintillators," US Patent, no. 4,421,671 (1983).
5. S. J. Duclos, Ch. D. Greskovich, R. J. Lyons, J. S. Vartuli, D. M. Hoffman, R. J. Riedner, M. J. Lynch, "Development of the HiLight™ scintillator for computed tomography medical imaging," *Nucl. Instr. Meth. Phys. Res. A*, **505** 68 (2003).
6. Suzuki *et al.*, "Radiation Detector," US Patent, no. 4,442,360 (1984).
7. Y. Ito, H. Yamada, M. Yoshida, H. Fujii, G. Toda, H. Takeuchi, Y. Tsukuda, "Hot Isostatic Pressed Gd_2O_2S:Pr, Ce, F Translucent Scintillator Ceramics for X-Ray Computed Tomography Detectors," *Jpn. J. Appl. Phys.*, **27** (1988) L1371.
8. H. Yamada, A. Suzuki, Y. Uchida, M. Yoshida, H. Yamamoto, "A Scintillator Gd_2O_2S:Pr, Ce, F for X-Ray Computed Tomography," *J. Electrochem. Soc.*, **136**, 2713 (1989).
9. Leppert *et al.*, "Method for producing a scintillator ceramic," US Patent, no. 5,296,163 (1994).
10. R. Hupke, C. Doubrava, "The new UFC-Detector for CT-Imaging," *Physica Medica*, **XV** N.4, 315 (1999).
11. E. Zych, C. Brecher, A. J. Wojtowicz, H. Lingertat, "Luminescence properties of Ce-activated YAG optical ceramic scintillator materials," *J. Lumin.*, **75**, 193 (1997).
12. A. Lempicki, C. Brecher, P. Szupryczynski, H. Lingertat, V.V. Nagarkar, S.V. Tipnis, S.R. Miller, "A new lutetia-based ceramic scintillator for X-ray imaging," *Nucl. Instr. Meth. Phys. Res. A*, **488**, 579 (2002).
13. Lempicki *et al.*, "High-density polycrystalline lutetium silicate materials activated with Ce," US Patent, no. 6,967,330 (2005).

14. H. S. Tripathi and V. K. Sarin, "Synthesis and densification of lutetium pyrosilicate from lutetia and silica," *Mat. Res. Bull.*, **42**, 197 (2007).

15. Y. Ji, D. Jiang, T. Fen, J. Shi, "Fabrication of transparent $La_2Hf_2O_7$ ceramics from combustion synthesized powders," *Mat. Res. Bull.*, **40**, 553 (2005).

16. Y. Ji, D. Jiang, J. Shi, "Preparation and spectroscopic properties of $La_2Hf_2O_7$:Tb," *Mat. Lett.*, **59**, 868 (2005).

17. Y. Ji, D. Jiang, J. Shi, "$La_2Hf_2O_7$:Ti ceramic scintillator for x-ray imaging," *J. Mater. Res.*, **20**, 567 (2005).

18. Xue-Jian Liu, Hui-Li Li, Rong-Jun Xie, Yi Zeng, Li-Ping Huang, "Cerium-doped lutetium aluminum garnet optically transparent ceramics fabricated by a sol-gel combustion process," *J. Mat. Res.*, **21**, 1519 (2006).

19. Y. Ji, D. Y. Jiang, J. J. Chen, L. S. Qin, Y. P. Xu, T. Feng, J. L. Shi, "Preparation, luminescence and sintering properties of Ce-doped $BaHfO_3$ phosphors," *Opt. Mat.*, **28**, 436 (2006).

20. S. L. Dole, S. Venkataramani, "Alkaline Earth Hafnate Phosphor with Cerium Luminescence," US Patent, no. 5,124,072 (1992).

21. V. S. Venkataramani, S. M. Loureiro, M. V. Rane, "Cerium-doped Alkaline-Earth Hafnium Oxide Scintillators having Improved Transparency and Method of Making Same," US Patent, no. 6,706,212 (2004).

22. A. V. Shevchenko, L. M. Lopato, G. I. Gerasimyuk, Z. A. Zaitseva, "Reactions in the system HfO_2 – SrO, HfO_2 – BaO, and ZrO_2 – BaO in high HfO_2 or ZrO_2 regions," *Izv. Akad. Nauk USSR, Neorg. Mat.*, **23**, 1495 (1987).

23. To be published, *J. Am. Ceram. Soc.*, ().

24. E.V.D. Van Loef, W.M. Higgins, J. Glodo, C. Brecher, A. Lempicki, V. Venkataramani, W.W. Moses, S.E. Derenzo, K.S. Shah, "Transparent Optical Ceramics for Scintillation Detection," in *Proceedings of the Glass & Optical Materials Division of the American Ceramic Society*, edited by

25. B.P. Mandal, Nandini Garg, Surinder M. Sharma, A.K. Tyagi, "Preparation, XRD and Raman spectroscopic studies on new compounds $RE_2Hf_2O_7$ (RE = Dy, Ho, Er, Tm, Lu, Y): Pyrochlores or defect-fluorite?," *J. Solid St. Chem.*, **179**, 1990 (2006).

26. A..V. Shevchenko, L.M. Lopato, I.E. Kir'yakova, "Interaction of HfO_2 with Y_2O_3, Ho_2O_3, Er_2O_3, Tm_2O_3, Yb_2O_3, and Lu_2O_3 at high temperatures," *Izv. Akad. Nauk SSSR, Neorg. Mat.*, **20**, 1991 (1984).

Mater. Res. Soc. Symp. Proc. Vol. 1038 © 2008 Materials Research Society

Energy Resolution and Non-proportionality of Scintillation Detectors

Marek Moszyński
Department of Detectors and Nuclear Electronics, Soltan Institute for Nuclear Studies, Swierk,
Otwock, PL 05-400, Poland

ABSTRACT

The limitation of energy resolution of scintillation detectors are discussed with a special emphasis on non-proportionality response of scintillators to gamma rays and electrons, which is of crucial importance to an intrinsic energy resolution of the crystals. Examples of the study carried out with different crystals and particularly those of tests of undoped NaI and CsI at liquid nitrogen temperature with the light readout by avalanche photodiodes are presented suggesting strongly that the non-proportionality of the halide crystals are not their intrinsic property. Moreover, the influence of slow components of the light pulses on energy resolution and non-proportionality are discussed.

INTRODUCTION

A γ-ray spectrometry with scintillation detectors belongs to the most important methods in the research and different applications of nuclear science. It covers, for example, a basic study of nuclear physics, environmental study, nuclear medicine and recently homeland security equipment. A great importance of scintillation detectors is associated with their high detection efficiency for nuclear radiation, capability to measure energy spectra, the possibility to work with a very high counting rate up to 10^7 counts/s and achievable best time resolution in coincidence or time-of-flight experiments. Capability to detect a wide assortment of radiations including γ and X-rays, charged particles and neutrons, the great variety in size and constitution of scintillators make them as the best choice in different applications [1].

For the γ-ray spectrometry the following properties of scintillation materials are essential [2]:

- A high density of the material and a high atomic number of the major element assuring high detection efficiency of γ-rays and a high photofraction,

- A high light output responsible for the high statistical accuracy of delivered signal,

- A fast decay time of the light pulse reflecting decay time of fluorescence components of the crystal, and allowing for a high counting rate measurements,

- A low contribution of the scintillator to the measured energy resolution associated mainly with its non-proportionality characteristics.

Three first properties are straightforward, as they are described by the basic properties of a scintillator. Energy resolution achievable with different crystals is the most mysterious. It is a function of the light output but it is also affected by internal properties of scintillator.

A good energy resolution is of the great importance for most of applications of scintillation detectors. Thus, its limitations are discussed below with a special emphasis on non-proportional response of scintillators to gamma rays and electrons, as it is of crucial importance to an intrinsic energy resolution of the crystals. An important influence of the scattering of secondary electrons (δ-rays) on intrinsic resolution is pointed out. Examples of the study carried out with different crystals, particularly those most important for γ-spectrometry are presented. The study of undoped NaI and CsI at liquid nitrogen temperature with the light readout by avalanche photodiodes suggests strongly that the non-proportionality of many crystals are not their intrinsic property and may be improved by a selective co-doping. Moreover, the influence of slow components of the light pulses on energy resolution and non-proportionality are discussed.

OUTLINE OF THE PROBLEM

The detection process of γ-rays in a scintillation detector can be described by a chain of subsequent processes, which introduce uncertainty in the measured energy, as a result of γ-rays absorbed in the detector. These processes can be identified as: 1) γ-ray absorption and light generation in the crystal, 2) light collection at the photocathode, 3) photoelectron production at the photocathode, 4) photoelectron collection at the first dynode and 5) multiplication by the PMT dynodes [2-4].

The energy resolution, $\Delta E/E$, of the full energy peak measured with a scintillator coupled to a photomultiplier can be written as:

$$(\Delta E/E)^2 = (\delta_{sc})^2 + (\delta_p)^2 + (\delta_{st})^2 \tag{1}$$

where δ_{sc} is the intrinsic resolution of the crystal, δ_p is the transfer resolution and δ_{st} is the PMT contribution to the resolution.

The statistical uncertainty of the signal from the PMT, corresponding to processes $3 - 5$, can be described, as:

$$\delta_{st} = 2.35 \times 1/N^{1/2} \times (1 + \varepsilon)^{1/2} \tag{2}$$

where N is the number of photoelectrons and ε is the variance of the electron multiplier gain, typically 0.1-0.2 for modern PMTs [5].

The PMT contribution can be determined experimentally based on the measured number of photoelectrons and it depends on the light output of the crystal being studied, quantum efficiency of the photocathode and efficiency of photoelectron collection at the first dynode.

The transfer component (processes $2 - 3$) is described by the variance associated with the probability that a photon from the scintillator results in the arrival of photoelectron at the first dynode and then is fully multiplied by the PMT. The transfer component depends on the quality of the optical coupling of the crystal and PMT, homogeneity of the quantum efficiency of the photocathode and efficiency of photoelectron collection at the first dynode. In the modern scintillation detectors the transfer component is negligible compared to the other components of the energy resolution [6].

The intrinsic resolution of the crystal (process 1) is connected mainly with the non-proportional response of the scintillator [4-25]. However, the experimentally determined intrinsic resolution is affected also by many effects such as inhomogeneities in the scintillator causing local variations in the light output and non-uniform reflectivity of the reflecting cover of the crystal. To discuss influence of the non-proportionality, one has to consider process of γ-ray absorption in the crystal.

A full-energy peak after gamma energy absorption results from electrons produced in the photoelectric absorption followed by emission and subsequent absorption of a cascade of X-rays and Auger electrons, and electrons generated by Compton scattering and terminated by photoelectric absorption. In the end the amount of light produced corresponding to the full energy deposition in the crystal of γ-quanta consists of contributions due to numerous secondary electrons that have a variety of energies. In the low energy region and in small volume crystals, photoelectric absorption dominates and the spread in the amount of light is due to different contributions from the X-ray and Auger electron cascade. At high energies, mainly in large volume crystals, Compton scattering is largely responsible for secondary electrons of different energies.

Another source of spread in the total light produced occurs when a given electron does not lose its energy in a unique manner in the crystal but produces further energetic electrons, known as δ-rays. In the low energy region, numerous low energy electrons, typically with energy below 10 keV [8] will affect energy resolution. In the high energy region scattered electrons have a higher energy and may more significantly influence the spread in the total light produced.

Several observations collected in the last 10 years on the influence of slow components of the light pulses on energy resolution suggest even more complex processes in the scintillators. This was done with CsI(Tl) [23,26], ZnSe(Te) [27], undoped NaI at liquid nitrogen temperature [28] and finally for NaI(Tl) at temperatures reduced below 0 °C [29]. A common conclusion of these observations is the fact that, in the case of scintillators showing two components of the light pulse decay, the highest energy resolution, and particularly the lowest contribution of the intrinsic resolution is obtainable when the spectrometry equipment integrates the whole scintillation light.

Although the intrinsic resolution of the scintillators appears to be mainly correlated with the non-proportional light response, some of the crystals, like LSO for example, seem to show a particularly poor energy resolution exceeding that expected from the non-proportionality [30]. In the study [30], a correlation of the non-proportionality and the intrinsic resolution of LSO crystals with their thermoluminescence integrated intensity and then with their afterglow were explored for the first time. In the recent study, the observed correlation of the intrinsic energy resolution of the LGSO crystals and the intensity of their afterglow suggests that the energy resolution of scintillation detectors may be affected also by a strong afterglow of the crystals [31,32].

STUDY OF ENERGY RESOLUTION AND NON-PROPORTIONALITY

One primary motivation and goal for a development of new scintillators is identifying a scintillator with the superior energy resolution. Consequently, it is important to understand the origins and relative contributions of the different energy resolution components. The studies were started already in fifties and a large number of papers were published to the end of sixties years discussing various aspects of the measured energy resolution, particularly for NaI(Tl) crystals [7-11,33-35]. The statistical contribution related to the light output and the photocathode sensitivity, influence of inhomogeneities of the light collected from various parts of the crystal and that of the variation of the photocathode quantum efficiency were discussed. But the major attention was paid to the influence of the non-proportional response of NaI(Tl). Prescott and Narayan [10-11] verified these mechanisms as the source of the intrinsic resolution.

Growing interest in the development of new scintillating detectors in the last decade has prompted efforts focused on developing a better understanding of the limitation of achievable energy resolution. Numerous studies have characterized the non-proportional response of new

Ce-doped scintillators [5,12-23]. It shows in general, a reduced light output at low energies. Dorenbos et al. [5,11] have discussed the effect of scintillator non-proportionality on the measured energy resolution in the light of new experimental data. Moszyński's group [2-4,6,19-25] have studied the non-proportionality and intrinsic resolution of numerous crystals using photomultipliers and avalanche photodiodes (APD).

Development of the Compton Coincidence Technique by Valentine and Rooney allowed a more accurate characterization of the electron response for a number of scintillators [13,18-24]. For the first time the true response for electrons was measured using a method not affected by the surface effects in scintillators. It confirms finally the non-proportional response of scintillators questioned by different experimenters [36].

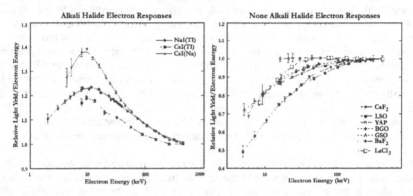

Figure 1. The relative scintillation response per unit energy deposited for fast electrons plotted as a function of energy for alkali halides (left figure) and non-alkali halides (right figure) crystals. The curves are normalized to unity at 445 keV. Note an excellent proportionality of YAP crystal. (Courtesy of W. Mengesha et al. [17])

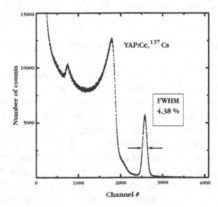

Figure 2. Energy spectrum of 661.6 keV γ-rays from a ^{137}Cs source measured with the YAP crystal, following [19].

Fig. 1 presents electron response of alkali halide crystals compared to that measured with non-alkali halides [17]. Instead of the light yield excess at low energies observed with alkali halide crystals a large reduction of the light yield was found in the second group of crystals. Only YAP is exceptional in that respect exhibiting nearly proportional response of the light yield.

Kapusta et al. [19] have shown that the proportional response of YAP results in a very good energy resolution of 4.38±0.11% for 662 keV γ-rays and also in the lowest observed intrinsic resolution of 1.3±0.5%, see Fig. 2. This was an important experimental fact showing clearly that the intrinsic resolution of scintillators is strongly correlated with the non-proportional response. It confirmed also a very small contribution of the transfer resolution component, included, in fact, to the determined intrinsic resolution of YAP crystal.

However, a Monte Carlo simulation of a contribution of the non-proportionality component to the intrinsic resolution, originating from the stopping process of γ-rays in NaI(Tl) and LSO crystals, done by Valentine et al. [18], could not explain the experimentally determined intrinsic resolution. It suggested that a contribution of a secondary electrons (δ-rays) to the intrinsic resolution is more important than estimated previously. Indeed, in the most recent papers [6,24-25], the importance of secondary electron scattering in γ-ray detection process on intrinsic resolution was pointed out.

In Ref. [6], the light output and energy resolution for the 10 mm in diameter and 10 mm high, and 75 mm in diameter and 75 mm high NaI(Tl) crystals were measured for γ-ray energies ranging from 16 keV to 1333 keV. These measurements enabled the observation of the light yield non-proportionality behavior and allowed the determination of the intrinsic resolution after correcting for the measured resolution for photomultiplier tube (PMT) statistics. The intrinsic resolution was then compared with the non-proportionality component calculated according to model of Ref. [18]. This comparison allowed the identification of the intrinsic resolution component associated with δ-rays, see Fig. 3. Consequently, it was shown that the δ-ray component is the most dominant component of the NaI(Tl) intrinsic resolution.

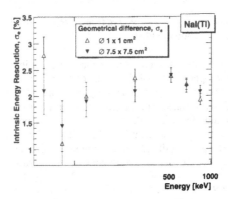

Figure 3. A δ-rays component determined for two NaI(Tl) crystals of different dimensions, following [6].

In Ref. [25], measurements carried out with small BGO crystals (∅9 mm × 5 mm) at LN$_2$ temperatures showed escape peaks distinct from photopeaks due to a good energy resolution of BGO and the energy of bismuth KX-rays of about 76 keV. Thus, a new idea, which appeared, was to analyze the escape peaks jointly with the full-energy peaks and compare their light outputs and energy resolutions.

Fig. 4 presents the example of the escape peak analysis of 320.1 keV γ-rays from a ^{51}Cr source [25]. The light yield and energy resolution of the main component of the escape peaks fit very well to those of the full energy peaks. Finally, the analysis of the energy resolution of the escape peaks in the energy range of 122 keV to 835 keV showed the good agreement between the intrinsic resolutions evaluated from the escape peaks in relation to the full energy peaks. This infers that the X-ray cascade, generated in the absorption process of γ-rays in scintillator, weakly affects the measured intrinsic resolution. This study seems to confirm finally that the scattering of electrons (δ-rays) is the most dominant component of the intrinsic resolution, as was concluded in a previous study [6].

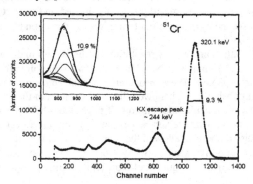

Figure 4. The energy spectrum of 320 keV γ-rays from a ^{51}Cr source. In the inset the fitting result of the escape peak is shown, following [25].

All the studies presented above seem to confirm that the intrinsic resolution of scintillators originating from the non-proportional response of the crystals is the fundamental limitation of obtainable energy resolution. It acts due to the absorption process of γ-rays and due to the scattering of electrons (δ-rays). But what is origin of the non-proportionality?

According to Murray and Meyer [28], the non-proportional response of the crystal to electrons was attributed to the fact that the probability of formation of an electron-hole pair depends on the specific energy loss, dE/dX, of the incident particle in the scintillator. The study carried out by Balcerzyk et al. [20-21] showed that the non-proportionality and the intrinsic resolution seem to depend on a structure of the crystals. The study of electron response of NaI(Tl), CsI(Tl) and CsI(Na) done by Mengesha et al. [17] leads to the same conclusion; the curves are similar in shape but different in magnitude (see Fig. 2). P.A. Rodnyi in his recent book „Physical processes in inorganic scintillators" has concluded that: „The non-proportionality of scintillator response is an intrinsic property of the (host) crystal and therefore cannot be improved substantially" [37]. This conclusion was supported by a good reproducibility of the measured non-proportionality characteristics for the crystals from different batches or manufacturers.

The study of the non-proportionality and intrinsic resolution of undoped oxide crystals as BGO, CWO and CaWO₄, carried out at both room and LN₂ temperatures, seems to confirm that in all cases the non-proportionality is a fundamental characteristic of undoped scintillator materials [25,38-39]. Fig. 5 presents the non-proportionality characteristic and the intrinsic resolution of a small BGO crystal measured at both room and LN₂ temperatures and. Note common curves independent of temperature. The points corresponding to the escape peaks fit very well to the curves.

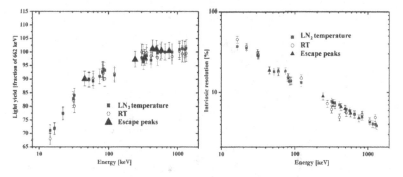

Figure 5. The non-proportionality characteristic and the intrinsic resolution of BGO crystal determined at both room and LN₂ temperatures. The points corresponding to the escape peaks, represented by the triangles, fit well within the curves, following [25].

In contrast, the recent studies of undoped NaI and CsI crystals at LN₂ temperature, using light readout by avalanche photodiodes, showed different non-proportionality curves for various tested samples, correlated with their purity [40-43].

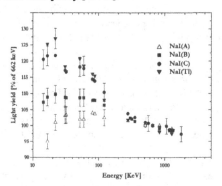

Figure 6. The non-proportionality curves of studied NaI crystals at LN₂ temperature in comparison to the curve measured for NaI(Tl) coupled to the XP2020Q photomultiplier, following [42].

Fig. 6 presents the non-proportionality curves of different samples of undoped NaI crystals, as measured at LN$_2$ temperature, in comparison to those of NaI(Tl) crystal at room temperature [41-42]. It is worth to note various curves for the different samples of the crystals correlated with the purity of the crystals, which was also reflected in the different emission spectra, see [42].

It suggests the question, is it possible to modify halide crystals, such as NaI(Tl) or CsI(Tl), to obtain a better non-proportionality and consequently, a better energy resolution?

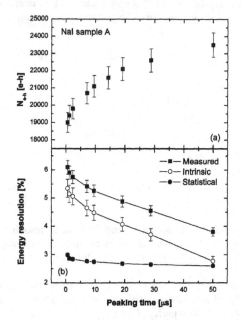

Figure 7. The measured energy resolution, statistical contribution, and calculated intrinsic energy resolution of NaI for 662 keV γ-peak versus amplifier peaking time at liquid nitrogen temperature. Also shown the yield of primary e-h pairs (upper panel). Following [41].

Several observations collected in the last 10 years on the influence of slow components of the light pulses on energy resolution suggest even more complex processes in the scintillators [32]. This was the most pronounced in case of one undoped NaI crystal. Fig. 7 presents dependence of the measured energy resolution, the statistical contribution and the calculated intrinsic resolution of NaI on the amplifier peaking time. The corrections for the statistical contribution of the number of e-h pairs on the peaking time were introduced using data presented in upper part of Fig. 7.

These results suggest that the observed performance is strongly related to the intrinsic resolution created in the tested NaI. A presence of the slow component, with the total intensity of about 20% up to 50 μs peaking time, is clearly seen in the upper panel of Fig. 7. However, its contribution weakly affects the statistical error of the energy resolution, see lower panel of Fig. 7. Note that the main component of the light pulse has the decay time constant below 100 ns [41]. Thus, the measured energy resolution is affected mainly by the intrinsic resolution. A similar effects were observed for CsI(Tl), ZnSe(Te) and NaI(Tl) at reduced temperatures below 0 °C.

Summarizing, all above discussed results suggest strongly that there is a correlation between energy resolution and non-proportionality measured at short and long shaping time constants. It is the most pronounced in the case of undoped NaI, which showed a dramatic improvement of both quantities due to the integration of the whole light. In the case of CsI(Tl), as well as NaI(Tl) at reduced temperatures, a weak amelioration of the non-proportionality observed at long shaping is reflected in a quite significant improvement of the intrinsic resolution for high energy gamma rays. It reflects importance of the non-proportionality of scintillator response on the contribution of the scintillator to its energy resolution.

The origin of the effect is not clear yet. It can be associated with a larger density of ionization for low energy secondary electrons, which may change the ratio of intensities of the fast to slow components of the light pulses. It follows well known faster decay time of light pulses for alpha particles in CsI(Tl), confirmed recently also for low-energy X-rays. A larger intensity of the fast component of the light pulses for 5.9 keV and 16.6 keV KX-rays than that measured for 662 keV γ-rays is reported in [44].

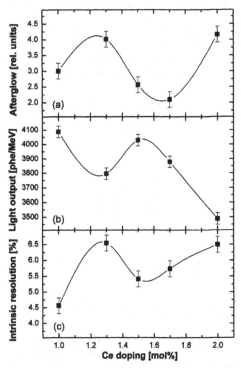

Figure 8. The intensity of the afterglow (a), the photoelectron number (b), and the intrinsic energy resolution (c), versus Ce doping of the LGSO crystals. Following [31].

Although the intrinsic resolution of the scintillators appears to be mainly correlated with the non-proportional light response, some of the crystals, like LSO for example, seem to show a poor energy resolution exceeding that expected from the non-proportionality [30]. Particularly, recently tested several samples of LGSO crystal showed different contribution of the intrinsic resolution in spite of common non-proportionality characteristics [31]. The observed correlation of the intrinsic energy resolution of the LGSO crystals and the intensity of their afterglow suggests that the energy resolution of scintillation detectors may be affected also by a strong afterglow of the crystals [31].

Fig. 8a shows the intensity of the afterglow versus doping of Ce in the studied LGSO crystals, while Figs 8b and 8c present corresponding photoelectron numbers and the intrinsic resolution.

A particularly interesting is correlation of the intrinsic energy resolution of the crystals with their afterglow. It could be related to the discussed above improvement of the energy resolution in the crystals, exhibiting two components of the light pulse decay. The difference would correspond only to the time scale of the effects, the afterglow of LGSO is in the second or hour's range, while the slow components discussed above are in the microsecond range. However, the observed effect seems to be different, as it does not affect the non-proportionality response of the crystals.

Recently invented LaCl$_3$ [45] and LaBr$_3$ [46] scintillators, characterizing by a high energy resolution, are of the great importance for gamma spectrometry. A very high energy resolution of about 2.7% for 662 keV gamma rays from a ^{137}Cs source, as measured with commercially available LaBr$_3$ crystals, is superior over all scintillation detectors.

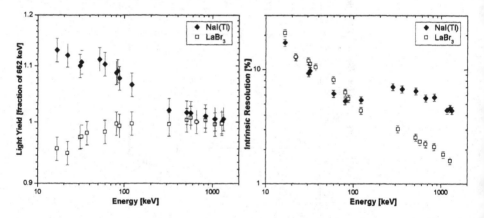

Figure 9. Non-proportionality of the light yield and intrinsic resolution of LaBr$_3$ and NaI(Tl).

An excellent energy resolution of LaBr$_3$ is due to its high light output and a good non-proportionality response of the crystal, shown in Fig. 9a, in comparison to NaI(Tl). It assures a low contribution of the intrinsic resolution to the measured energy resolution, dramatically lower than that of NaI(Tl), as shown in Fig. 9b.

DISCUSSION AND CONCLUSIONS

In spite of large efforts done by many groups, the non-proportionality of the light yield and the intrinsic resolution of the scintillation crystals remain far from being completely understood. The collected numerous observations concerning non-proportionality and intrinsic resolution of scintillators can be summarized, as follow:

- The non-proportionality is the fundamental limitation of energy resolution. It acts by numerous secondary γ and X-ray quanta, as well as, by a scattering of secondary electrons (δ-rays) in the absorption process of γ-rays in the crystal. The last process seems to be dominating in creation of the intrinsic resolution. It is confirmed by a high energy resolution of YAP, LaCl$_3$ and LaBr$_3$, which show a low non-proportionality of the light output in a large range of γ-ray energy.
- The non-proportionality of the undoped oxide crystals as BGO, CWO and CaWO$_4$ seems to be a fundamental characteristic of scintillator material.
- Study of pure undoped halide crystals, as NaI and CsI, suggests that their non-proportionality and intrinsic resolution characteristics are altered by accidental doping by impurities. This observation suggests that a selective co-doping of crystals may improve the non-proportionality and the intrinsic resolution.
- The observed correlation between the energy resolution and non-proportionality measured at short and long shaping time constants and finally expected influence of afterglow on energy resolution suggest even more complex processes in the scintillators [32].

Recently, Dorenbos [48] proposed to analyze non-proportionality in terms of two processes. The first one is related to the non-proportionality of the host material, while the second process is associated with the transport of the energy to the activator. Both processes are in fact correlated with the ionization density, as it was postulated by Murray and Meyer [35].

This approach seems to be confirmed by the recent experimental studies. The non-proportionality of the undoped oxide crystals, as BGO [25] and CWO [38], represents the fundamental properties of the scintillator materials. In case of doped crystals, one observes variation of the non-proportionality characteristic depending on doping agents, particularly for halide crystals [42-43]. Moreover, the pure, undoped halide crystals, as NaI [42] and CsI [43], are very sensitive to the impurities, which affect the non-proportionality curves [42-43]. It suggests that a selective co-doping of scintillators may improve the non-proportionality and their energy resolution [43]. No doubt that further studies are necessary.

ACKNOWLEDGMENTS

The author would like to thank Dr. A. Syntfeld-Każuch for discussion and critical review of the manuscript.

REFERENCES

1. G.F. Knoll, "Radiation Detection and Measurements", Third Edition, John Willey and Son, New York, 2000.
2. M. Moszyński, *Nucl. Instr. Meth.*, **A505**, 101 (2003).
3. M. Moszyński, *SPIE Proceedings*, **Vol. 5922**, 592205-1.

4. M. Moszyński, "Non-proportionality and energy resolution of scintillation detectors" in *Radiation Detectors for Medical Imaging*, Ed. S. Tavernier, A. Gektin, B. Grinyov, W.W. Moses, Springer, Dordrecht, The Netherlands(2006)

5. P. Dorenbos, J.T.M. de Haas, C.W.E. van Eijk, , *IEEE Trans. Nucl. Sci.*, **42**, 2190 (1995).

6. M. Moszyński, J. Zalipska, M. Balcerzyk, M. Kapusta, W. Mengeshe, J.D. Valentine, *Nucl. Instr. Meth. A*, **A484**, 259 (2002).

7. G.G. Kelly, P.R. Bell, R.C. Davis and N.A. Lazar, *IRE Trans. on Nucl, Sci.*, NS3, 57 (1956).

8. P. Iredale, *Nucl. Instr. Meth.*, **11**, 340 (1961).

9. C.D. Zerby, A. Meyer, R.B. Murray, *Nucl. Instr. Meth.*, **12**, 115 (1961).

10. G.H. Narayan and J.R. Prescott, *IEEE Trans. Nucl. Sci.*, NS-15, 162 (1968).

11. J.R. Prescott and G.H. Narayan, *Nucl. Instr. Meth.*, **75**, 51 (1969).

12. P. Dorenbos, J.T.M. de Hass, C.W.E. van Eijk, C.L. Melcher, and J.S. Schweitzer, *IEEE Trans. Nucl. Sci.*, **41**, 1052 (1994).

13. T.D. Taulbee, B.D. Rooney, W. Mengesha, and J.D. Valentine, *IEEE Trans. Nucl. Sci.*, **44**, 489 (1997).

14. J.D. Valentine and B.D. Rooney, *Nucl. Instr. Meth.*, **A353**, 37 (1994).

15. B.D. Rooney and J.D. Valentine, *IEEE Trans. Nucl. Sci.*, **43**, 1271 (1996).

16. B.D. Rooney and J. D. Valentine, *IEEE Trans. Nucl. Sci.*, **44**, 509 (1997).

17. W. Mengesha, T.D. Taulbee, B.D. Rooney, and J.D. Valentine, *IEEE Trans. Nucl. Sci.*, **45**, 456 (1998).

18. J.D. Valentine, B.D. Rooney, and J. Li, *IEEE Trans. Nucl. Sci.*, **45**, 512 (1998).

19. M. Kapusta, M. Balcerzyk, M. Moszyński, J. Pawelke. *Nucl. Instr. Meth.*, **A421**, 610 (1999).

20. M. Balcerzyk, M. Moszyński, M. Kapusta, *Proc. of Fifth Int. Conf. on Inorganic Scintillators and Their Application*, Moscow, August 16-20, 167 (1999).

21. M. Balcerzyk, M. Moszyński, M. Kapusta, D. Wolski, J. Pawelke, C.L. Melcher, *IEEE Trans. Nucl. Sci.*, **47**, 1319 (2000).

22. M. Moszyński, M. Kapusta, D. Wolski, M. Szawlowski, W. Klamra, *IEEE Trans. Nucl. Sci.*, **45**, 472 (1998).

23. M. Moszyński, M. Kapusta, J. Zalipska, M. Balcerzyk, D. Wolski, M. Szawlowski, and W. Klamra, *IEEE Trans. Nucl. Sci.*, **46**, 243 (1999).

24. W. Mengesha, J.D. Valentine, *IEEE Trans. Nucl. Sci.*, **49**, 2420 (2002).

25. M. Moszyński, M. Balcerzyk, W Czarnacki, M. Kapusta, W. Klamra, A. Syntfeld, M. Szawlowski, *IEEE Trans. Nucl. Sci.*, **51**, 1074 (2004).

26. A. Syntfeld, Ł. Świderski, W. Czarnacki, M. Moszyński, W. Klamra, *IEEE Trans. Nucl. Sci.*, in press.

27. M. Balcerzyk, W. Klamra, M. Moszyński, M. Kapusta, M. Szawlowski, *Nucl. Instr. Meth. A*, **A482**, 720 (2002).

28. M. Moszyński, W. Czarnacki, W. Klamra, M. Szawlowski, P. Schotanus, M. Kapusta, *Nucl. Instr. Meth. A*, **A505**, 63 (2003).

29. Ł. Świderski, M. Moszyński, W Czarnacki, A. Syntfeld-Każuch, M. Gierlik, „ *IEEE Trans. Nucl. Sci.*, **54**, 1372 (2007).

30. M. Kapusta, P Szupryczyński, C. Melcher, M. Moszyński, M. Balcerzyk, A. A. Carey, W. Czarnacki, A. Spurrier, A. Syntfeld, *IEEE Trans. Nucl. Sci.*, **52**, 1098 (2005).

31. M. Moszyński, A. Nassalski, W. Czarnacki, A. Syntfeld-Każuch, D. Wolski, T. Batsch, T. Usui, S. Shimizu, N. Shimura, K. Kurashige, K. Kurata, and H. Ishibashi, *IEEE Trans. Nucl. Sci.*, **54**, 725 (2007).

32. M. Moszynski, A. Nassalski, A. Syntfeld-Każuch, Ł. Swiderski, T. Szczęśniak, *IEEE Trans. Nucl. Sci.*, in press.
33. E. Breitenberger, *Progr. Nucl. Phys.* **4**, 56(1956).
34. D. Engelkemeir, *Rev. Sci. Instr.*, **27**, 589 (1956).
35. R.B. Murray and A. Meyer, *Phys. Rev.*, **122**, 815 (1961).
36. H. Leutz, C. D'Ambrosio, *IEEE Trans. Nucl. Sci.*, **44**, 190 (1997).
37. P.A. Rodnyi, "Physical processes in inorganic scintillators", CRC Press, New York, 1997.
38. M. Moszyński, M. Balcerzyk, M. Kapusta, A. Syntfeld, D. Wolski, G. Pausch, J. Stein, P. Schotanus, *IEEE Trans. Nucl. Sci.*, **52**, 3124 (2005).
39. M. Moszyński, M. Balcerzyk, W. Czarnacki, A. Nassalski, T. Szczęśniak, M. Moszyński, H.Kraus, V. B. Mikhailik, I.M. Solskii, *Nucl. Instr. Meth. A* **A553**, 578 (2005).
40. M. Moszyński, W. Czarnacki, M. Kapusta, M. Szawlowski, W. Klamra, P. Schotanus, *Nucl. Instr. Meth. A.*, **A486**, 13 (2002).
41. M. Moszyński, W. Czarnacki, W. Klamra, M. Szawlowski, P. Schotanus, M. Kapusta, *Nucl. Instr. Meth. A*, **A505**, 63 (2003).
42. M. Moszynski, M. Balcerzyk, W. Czarnacki, M. Kapusta, W. Klamra, P. Schotanus, A. Syntfeld, M. Szawlowski, *IEEE Trans. Nucl. Sci.*, **50**, 767 (2003).
43. M. Moszyński, M. Balcerzyk, W. Czarnacki, M. Kapusta, W. Klamra, P. Schotanus, A. Syntfeld, M. Szawlowski, *Nucl. Instr. Meth. A*, **A537**, 357(2005).
44. A. Syntfeld-Każuch, M. Moszyński, Ł. Świderski, W. Klamra, A. Nassalski, IEEE Trans. Nucl. Sci., in press.
45. E.V.D. van Loef, P. Dorenbos, C.W.E. van Eijk, *Appl. Phys. Lett.*, **77**, 1467 (2000).
46. E.V.D. van Loef, P. Dorenbos, C.W.E. van Eijk, K. Kramer, H.U. Gudel, *Appl. Phys. Lett.*, **79**, 1573 (2001).
47. P. Dorenbos, "Thoughts on non-proportionality", *Non-proportionality Workshop*, Portland, USA, October 2003.

Mater. Res. Soc. Symp. Proc. Vol. 1038 © 2008 Materials Research Society   1038-O08-01

Passivation of Semiconductor Surfaces for Improved Radiation Detectors: X-ray Photoemission Analysis

A J Nelson[1], A M Conway[2], C E Reinhardt[2], J L Ferreira[1], R J Nikolic[2], and S A Payne[1]

[1]MSTD, LLNL, 7000 East Avenue, Livermore, CA, 94550
[2]ENG, LLNL, 7000 East Avenue, Livermore, CA, 94550

ABSTRACT

Surface passivation of device-grade radiation detector materials was investigated using x-ray photoelectron spectroscopy in combination with transport property measurements before and after various chemical treatments. Specifically Br-MeOH (2% Br), KOH with NH_4F/H_2O_2 and NH_4OH solutions were used to etch, reduce and oxidize the surface of unoriented $Cd_{(1-x)}Zn_xTe$ semiconductor crystals. Scanning electron microscopy was used to evaluate the resultant microscopic surface morphology. Angle-resolved high-resolution photoemission measurements on the valence band electronic structure and core lines were used to evaluate the surface chemistry of the chemically treated surfaces. Metal overlayers were then deposited on these chemically treated surfaces and the I-V characteristics measured. The measurements were correlated to understand the effect of interface chemistry on the electronic structure at these interfaces with the goal of optimizing the Schottky barrier height for improved radiation detector devices.

INTRODUCTION

The development of cadmium zinc telluride ($Cd_{(1-x)}Zn_xTe$, CZT) as a nuclear radiation detector material has progressed with advances in CZT crystal growth that minimizes bulk defects and precipitates.[1,2] With this progress, our focus is now on surface properties of CZT since the interfacial chemistry has a powerful influence on the electrical stability of ohmic and Schottky contacts to CZT. The performance of CZT as a room temperature radiation detector can thus be improved with careful attention to modification of the surface chemistry. Chemical treatments of the CZT surface prior to application of electrical contacts require comprehensive characterization to elucidate advantageous changes in surface electronic structure.

Mechanical polishing followed by bromine-based etching is routinely employed for CZT surface preparation prior to device fabrication. This treatment removes the native oxide but leaves a Te-rich surface resulting in metal/CZT interface degradation and excessive leakage currents. [1-3] Alternative surface preparation methods have been proposed for surface passivation [3-6] but have not been fully characterized nor implemented for practical device fabrication. A novel two-step chemical passivation process for CZT was reported in [4]. This paper further characterizes this chemical process in terms of the surface chemistry, surface morphology and electronic structure of chemically treated $Cd_{(1-x)}Zn_xTe$ and correlates the results with transport properties.

EXPERIMENT

Device-grade unoriented p-$Cd_{(1-x)}Zn_xTe$ was polished and subjected to various chemical treatments. Initially a 2 min Br-MeOH (2% Br) etch was performed and the resultant surface chemistry was characterized by x-ray photoelectron spectroscopy (XPS). Alternatively, following Br:MeOH etching, the CZT was treated with KOH (15 % in water) for 40 min, rinsed with DI water, then subjected to NH_4F/H_2O_2 (10%/10% in water) for 45 min, rinsed again with DI water and blown dry. Concentrated NH_4OH was also used to treat the oxidized CZT surface for 2 min at ambient temperature.

XPS analysis was performed on a PHI Quantum 2000 system using a focused monochromatic Al Kα x-ray (1486.7 eV) source for excitation and a spherical section analyzer. The instrument has a 16-element multichannel detection system. A 200 μm diameter x-ray beam was used for analysis. The x-ray beam is incident normal to the sample and the x-ray detector is at 45° away from the normal. The pass energy was 23.5 eV giving an energy resolution of 0.3 eV that when combined with the 0.85 eV full width at half maximum (FWHM) Al Ka line width gives a resolvable XPS peak width of 1.2 eV FWHM. Deconvolution of non-resolved peaks was accomplished using Multipak 6.1A (PHI) curve fitting routines. The collected data were referenced to an energy scale with binding energies for Cu $2p_{3/2}$ at 932.72± 0.05 eV and Au $4f_{7/2}$ at 84.01± 0.05 eV. Binding energies were also referenced to the C 1s photoelectron line arising from adventitious carbon at 284.8 eV. Low energy electrons and argon ions were used for specimen neutralization.

Diodes were fabricated by depositing platinum top contacts with guard rings and a gold backside contact. Current versus voltage measurements were performed on these diodes to determine the effect of contact interface chemistry on detector transport properties.

RESULTS AND DISCUSSION

XPS survey spectra of the etched, reduced and oxidized surfaces of p-$Cd_{(1-x)}Zn_xTe$ were acquired to determine surface stoichiometry and impurity concentrations. The quantitative surface compositional analyses and elemental ratios are summarized in Table I. Cd/Te ratio indicates that (1) 2%Br:MeOH etch results in a Te-rich surface, (2) KOH + NH_4F/H_2O_2 treatment results in a stoichiometric surface, and (3) NH_4OH results in a Te-rich surface.

Table I. Relative XPS Surface Compositional Analysis (atomic %) of the Chemically Treated $Cd_{(1-x)}Zn_xTe$

Sample	Cd	Zn	Te	Cd/Te
CZT, as received	50.7	2.0	47.3	1.07
CZT – 2%Br:MeOH etch	31.6	2.0	66.4	0.48
CZT - KOH + NH_4F/H_2O_2 treatment	50.5	-	49.5	1.02
CZT - NH_4OH treatment	40.9	1.7	57.4	0.71

High magnification planar view scanning electron microscopy (SEM) photomicrographs are presented in Figure 1(a) and (b), respectively, of the Br:MeOH etched and KOH + NH_4F/H_2O_2 treated CZT surfaces. The observed morphology on the surface of the Br:MeOH etched CZT has an orange peel texture. The surface morphology of the KOH + NH_4F/H_2O_2

treated CZT surface shows some pitting in addition to the formation of a thin surface oxide (≈10 nm).

Figure 1. SEM photomicrographs of (a) Br:MeOH etched CZT and (b) KOH + NH₄F/H₂O₂ treated CZT.

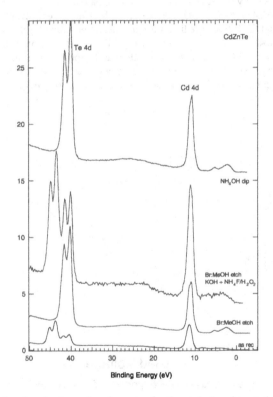

Figure 2. XPS core-level and valence band spectra for the etched and treated CZT surfaces.

Figure 2 shows the shallow core-levels and valence band region for the as received, Br:MeOH etched, KOH + NH_4F/H_2O_2 treated, and NH_4OH etched CZT surfaces. These spectra contain the Te 4d, Cd 4d and upper valence band, and provide unique information about the electronic structure and the nature of chemical bonding at the CZT surface. The evolution of the electronic structure at the surface is indicated by the transition of the valence band maximum (VBM).

In these spectra, the Te $4d_{5/2,3/2}$ core-level has two spin-orbit pairs. The higher binding energy spin-orbit pair at 43.8 eV and 45.0 eV represent an oxide ($TeO_2/CdTeO_3$) and the lower binding energy pair at 40.6 eV and 41.8 eV represents Te in CZT in agreement with literature values. [7] The Cd 4d spin-orbit pair cannot be resolved with our instrumentation, so the measured centroid of this peak is at 11.4 eV. In addition, the full width half maximum (FWHM) of the Cd 4d peak indicates the presence of a surface oxide ($CdTeO_3$) for as received CZT. Following the Br:MeOH etch the Te 4d oxide components disappear indicating removal of the native oxide. Also, Te $4d_{5/2,3/2}$ the spin-orbit pair and the Cd 4d peak shifts 0.4 eV to lower binding energy. In addition, the intensity of the Te 4d peaks increases relative to the Cd 4d intensity further supporting the presence of a Te-rich surface, possibly with the formation of Te islands. Formation of Te islands would account for the increased Te signal relative to the Cd signal due to the visibility of the substrate between the islands.

Following KOH + NH_4F/H_2O_2 treatment, the Te $4d_{5/2,3/2}$ core-level spectra show two component pairs and the Cd 4d core line broadens indicating oxide formation. The energy shifts observed for the Te $4d_{5/2,3/2}$ spin-orbit components following this peroxide treatment is similar to those observed for the Cd 4d core level. Also note that the relative intensity ratio of the Te 4d and Cd 4d has change. Recall that compositional analysis determined this to be a stoichiometric surface. Electrical contacts are deposited on this oxide surface for the following transport property measurements.

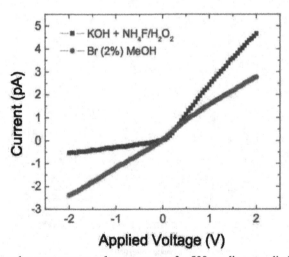

Figure 3. Measured current versus voltage response for 500μm diameter diodes with guard rings.

The final surface chemical treatment we explored was a NH₄OH dip at ambient temperature for 2 min. The resulting Te 4d and Cd 4d spectrum reveals that the oxide was removed by this treatment. Again note the change in the relative intensity ratio of the Te 4d and Cd 4d supporting the compositional analysis that indicated this is a Te rich surface. This treatment could be a replacement for the standard Br:MeOH etch for future CZT device fabrication.

Measured current versus voltage curves for 500 μm diameter diodes with guard rings are shown in figure 3. Rectifying behavior was achieved for the device treated with the KOH + NH₄F/H₂O₂ solution while the Br:MeOH sample shows ohmic behavior. We infer that the oxide layer created by the KOH + NH₄F/H₂O₂ solution is blocking current flow from the Pt contact into the semiconductor.

CONCLUSIONS

X-ray photoelectron spectroscopy has been used to determine the effects of wet chemical etching/treatment on the surface chemistry and surface electronic structure of CZT. Results show that 2% Br:MeOH removes the surface oxide and that the KOH + NH₄F/H₂O₂ treatment yields a well-behaved oxide surface. The NH₄OH treatment resulted in an oxide free surface and may be a preferred surface preparation treatment for CZT devices. The I-V characteristics reveal that KOH + NH₄F/H₂O₂ treatment produces rectifying characteristics while the 2% Br:MeOH treatment results in ohmic behavior. Future work will include treating CZT with (NH₄)₂S and NH₄OH/thiourea solutions and measuring I-V and C-V characteristics.

ACKNOWLEDGMENTS

This work performed under the auspices of the U.S. Department of Energy by Lawrence Livermore National Laboratory under Contract DE-AC52-07NA27344.

REFERENCES

1. T.E. Schlesinger, J.E. Toney, H. Yoon, E.Y. Lee, B.A. Brunett, L. Franks, and R.B. James, Materials Sci. Eng. 32, 103 (2001).
2. T. Takahashi and S. Watanabe, IEEE Trans. Nucl. Sci. 48(4), 950 (2001).
3. K.-T. Chen, D. T. Shi, H. Chen, B. Granderson, M. A. George, W. E. Collins, and A. Burger, J. Vac. Sci. Technol. A15(3), 850 (1997).
4. S. Wenbin, W. Kunshu, M. Jiahua, T. Jianyong, Z. Qi and Q. Yongbiao, Semicond. Sci. Technol. 20 , 343 (2005).
5. V. G. Ivanitska, P. Moravec, J. Franc, Z. F. Tomashik, P. I. Feychuk, V. M. Tomashik, L. P. Shcherbak, K. Masek, and P. Hoschl, J. Electron. Mater. 36(8), 1021 (2007).
6. L. Qiang and J. Wanqi, Surface states and passivation of p-Cd0.9Zn0.1Te crystal, Nucl. Instrum. Methods A 562(1), 468 (2006).
7. P. John, T. Miller, T.C. Hsieh, A.P. Shapiro, A.L. Wachs, and T.-C. Chiang, Phys. Rev. B34, 6704 (1986).

Mater. Res. Soc. Symp. Proc. Vol. 1038 © 2008 Materials Research Society 1038-O08-04

Lutetium Oxide Coatings by PVD

Stephen G Topping, C H Park, S K Rangan, and V K Sarin
Department of Manufacturing Engineering, Boston University, 15 St. Mary's Street, Boston, MA, 02215

ABSTRACT

Due to its high density and cubic structure, Lutetium oxide (Lu_2O_3) has been extensively researched for scintillating applications. Present manufacturing methods, such as hot pressing and sintering, do not provide adequate resolution due to light scattering of polycrystalline materials. Vapor deposition has been investigated as an alternative manufacturing method. Lutetium oxide transparent optical coatings by magnetron sputtering offer a means of tailoring the coating for optimum scintillation and resolution. Sputter deposited coatings typically have inherent stress and defects that adversely affect transparency and emission. The effect of process parameters on the coating properties is being investigated via x-ray diffraction (XRD), scanning electron microscopy (SEM) and emission spectroscopy, and will be presented and discussed.

INTRODUCTION

Rare earth oxides have been extensively used in the x-ray detector industry for quite some time due to their stability, high density and high atomic number [1]. However, they have generally been limited to small area detectors due to manufacturing limitations. Lutetium Oxide (Lu_2O_3) doped with Europium Oxide (Eu_2O_3) has been studied using hot pressing and sintering as an alternative to the industry standard Cesium Iodide doped with Tantalum (CsI:Tl). In terms of optical and scintillating properties, CsI:Tl has a good transparency, density of 4.51g/cc and emits ~60,000 photons per MeV of incident x-rays [2]. Compared to $Lu_2O_3:Eu^{3+}$, which has a highly transparent BCC structure, a density of 9.4g/cc and emits ~30,000 photons per MeV [1]. High density and high atomic number of Lu_2O_3 makes it an ideal scintillator. A viable manufacturing process would expand its market to the large area scintillators.

Current manufacturing methods, such as sintering and hot pressing produce a transparent 2-3mm thick disc that must be ground and polished to a thickness close to the desired thickness. It must then be pixelized into 20µm by 20µm square pixels as shown in Figure 1 using a highly labor intensive laser ablation process to reduce light scattering. The top surface is then placed on the CCD camera using optical glue and the back is ground off. [2] Dentistry is one of the applications for such a device and requires approximately 200 microns of Lu_2O_3, compared to 2mm for CsI, to absorb most of the incoming x-rays. Our proposal is to develop vapor deposited Lu_2O_3 coatings as an alternative manufacturing method that would enable largescale detector fabrication.

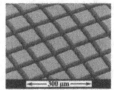

Figure 1. Top surface scanning electron image of a laser pixelized Lutetium Oxide ceramic. [2]

EXPERIMENT

The radio frequency (R.F.) magnetron sputtering setup used had a 2 inch diameter target angled at 45 degrees with respect to the substrate. The target was made by hot pressing Lu_2O_3 powder doped with 5 mol% Eu_2O_3 at 1700°C using a graphite uniaxial hot press. A thin 2 inch diameter graphite disc was used as the substrate and it was rotated at approximately 20rpm to increase uniformity. The R.F. power source was an Advanced Energy RFX600 capable of producing 600 Watts.

Coatings were deposited at 50, 75 and 100 Watts and examined. It was determined that 100 Watts was the maximum useable power level, above which charging and target damage occurs. The coatings were examined using a Bruker D8 Focus X-ray diffraction (XRD) unit using Cu-Kα radiation to determine orientation, and a Zeiss field emission scanning electron microscope (SEM) to examine the microstructure. All coatings were heat post-treated in a Tungsten furnace at 900°C in an argon atmosphere for 2 hours.

DISCUSSION

Microstructural analysis of the top surface and the fractured cross sections, as shown in Figure 2, revealed a strong morphological dependence on input power. The surface images showed a clear transition from what appears to be cellular growth to plate growth. We are further investigating the growth using transmission electron microscopy (TEM) to fully characterize the growth mechanism. At 50W and 100W the columnar growth appears to be of uniform width and perpendicular to the surface, whereas at 75W, the columnar growth becomes radial. The diameter of the columnar growth is not clear from the fractured cross section. However, with top surface images in figure 2, clearly show larger boundaries, indicative of a columnar grain growth. As expected, grain diameter measurements indicated a trend of decreasing columnar diameter with increasing power (or deposition rate) as shown in Table 1.

The columnar growth was determined to be (100) textured for low input power and (111) textured for high input power as determined from the XRD pattern (Figure 3). It is noteworthy that the intensities of the (100) and (111) peaks are low, indicating that crystallinity/orientation is not significant. Low intensity diffraction peaks from other planes further suggest that not all growths are perpendicular and potentially slightly polycrystalline. This is most likely a result of slow kinetics because the low thermal energy does not enable the newly formed grains to grow epitaxially. Furthermore, all the XRD patterns are increasingly shifted towards a larger unit cell with increasing power, which is a typically attributed to growth stresses. The <100> is a lower energy growth direction and with sufficient stresses can induce a shift towards <111> growth.

In a PVD sputtering system, the plasma intensity is dependent on the power applied, which also affects the sputtering rate. The plasma itself can attain high temperatures and can provide some thermal energy to the coating and the substrate can reach temperatures up to 100°C. However, the plasma provides a relatively large amount of thermal energy to a very thin layer, notably the deposition layer. This is believed to be the reason for the drastic change in coating morphology observed at 75W. At this power there is a balance between deposition rate and thermal energy provided by the plasma that enables better crystallization. At 50W the low intensity plasma provides low thermal energy and despite reduced deposition rates, is not adequate for crystalline growth. At 100W, despite increased plasma thermal energy, the atoms do not have sufficient time to rearrange because of the higher density of incoming atoms.

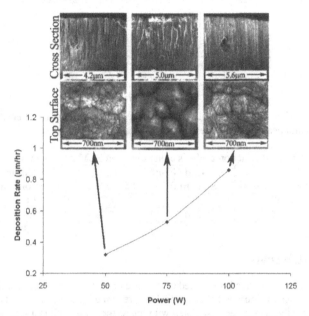

Figure 2. Effect of power on coating morphology and growth rates. The fractured cross section images and their respective surface morphologies have been shown.

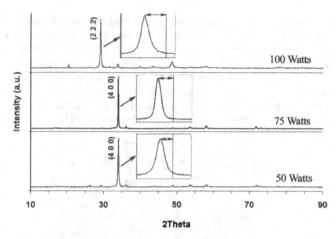

Figure 3. X-ray diffraction patterns of the as deposited coatings. Highest intensity peaks have been magnified and peak shift has been emphasized.

Table I. X-ray diffraction analysis compared with SEM grain size measurements

Power (Watts)	Measured Grain Size (nm)	Lattice Distortion	Volume Distortion
50	415	1.0%	3.0%
75	290	1.2%	3.6%
100	247	2.1%	6.3%

Heat Treatment Analysis

The samples were then heat treated to increase crystallinity and observe changes in morphology. As seen in figure 4, the (100) peak has reverted back to the theoretical position indicating stress relief. However, associated with the restored unit cell is a subsequent volume change resulting in reduced thickness and cracking (Figure 4). In the 100W case, the volume distortion leads to loss of adhesion making further analysis on the heat treated sample almost impossible. One can observe in Figure 4 that the morphology of the coating remains identical to the as-deposited coating indicating the coating stability. A small increase in the intensity of the (100) peak was observed indicating slight grain growth or increase in crystallinity. In figure 5 it can be seen that the edge of the 100W sample remains adherent, which can be attributed to the non-uniformity of deposition conditions. In magnetron sputtering, a ring source is created which in our case is angled at approximately 45 degrees to a rotating substrate. The angling and rotation is used to improve thickness uniformity but results in non-uniform plasma heating and deposition angles which are critical growth factors. Furthermore, the kinetic energy of the ejected material plays a crucial role in the coating properties and is a function of the travel distance and total pressure. Therefore, the center of the substrate will be exposed to relatively constant deposition conditions, whilst the outer edges will vary significantly every half rotation. XRD pattern of the outer edge is that of a partially polycrystalline coating.

118

One of the indicators of the extent of crystallization in a scintillating material is the emission intensity and spectrum. The emission spectrum of the 'as deposited' and heat treated samples were measured using cathodoluminescence. The 'as deposited' emission intensity was found to be too low to be detected whilst the heat-treated samples appeared to have a standard emission spectrum. Ultraviolet light at 254 nm also induces emission due to the charge transfer band at approximately 250nm in the host material [4] as seen in Figure 5. The lack of emission can be attributed to either low crystallinity or a large number of defects that act as charge traps resulting in non-radiative transitions. Once heat-treated, the defects are mostly eliminated and increased crystallinity results in improved emission. The 75W sample produced the highest emission intensity, further confirming the previously reported results.

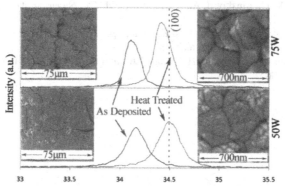

Figure 4. X-ray diffraction pattern of the (100) peak to emphasize stress reduction. On the left: low magnification image of the top surface. On the right: high magnification image of the top surface.

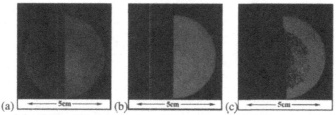

Figure 5. Ultraviolet (254nm) light excitation of the coatings. Left half: as deposited. Right half: heat treated. Input power: (a) 50W (b) 75W (c) 100W.

CONCLUSIONS

RF magnetron sputtered coatings of $Lu_2O_3:Eu^{3+}$ were successfully deposited using vapor deposition. The as-deposited coatings were partially crystalline and did not scintillate. Thermal treatment of the coatings resulted in increased crystallinity and lower defects, leading to excellent scintillation. The columnar nature of the coatings potentially makes this a very attractive candidate for use in x-ray imaging, eliminating the need for the highly labor intensive laser pixelization process.

ACKNOWLEDGMENTS

This research has been partially supported by the National Institute of Health under grant No. 5R21EB005037.

REFERENCES

1. E. Zych, *J. Phys. Condens. Matter* **14**, 5637 (2002).
2. I. Shestakova, V. Gaysinskiy, J. Antal, L. Bobek and V.V. Nagarkar, *Nucl. Instr. and Meth. In Phys. Res.* B **263**, 234 (2007).
3. T. Igarashi, M. Ihara, T. Kusunoki and K. Ohno, *App. Phys. Let.* **76** [12], (2000).
4. J. Trojan-Piezga, E. Zych, D. Hreniak and W. Strek, *J. of Alloys and Comp.* **380**, 123 (2004).
5. Y. Satoh, H. Najafov, S. Ohshio, H. Saitoh, *Advances in Tech. of Mat. And Mat. Processing*, **17** [1], 43-46 (2005).

Effects of Gamma Irradiation on Optical Properties of Colloidal Nanocrystals

Nathan J. Withers[1], Krishnaprasad Sankar[1], Brian A. Akins[1], Tosifa A. Memon[1], Jiangjiang Gu[1,2], Tingyi Gu[1,2], Shin T. Bowers[1,3], Melisa R. Greenberg[1,4], Gennady A. Smolyakov[1], Robert D. Busch[5], and Marek Osinski[1]

[1]Center for High Technology Materials, University of New Mexico, 1313 Goddard SE, Albuquerque, NM, 87106-4343

[2]School of Electronic, Information, and Electrical Engineering, Shanghai Jiao Tong University, Shanghai, China, People's Republic of

[3]Division of Engineering, Brown University, Box 0424, Providence, RI, 02912

[4]CVI Laser, LLC, 200 Dorado Pl. SE, Albuquerque, NM, 87123

[5]Department of Chemical and Nuclear Engineering, University of New Mexico, Albuquerque, NM, 87131

ABSTRACT

The effects of ^{137}Cs gamma irradiation on photoluminescence properties, such as spectra, light output, and lifetime, of several types of colloidal nanocrystals have been investigated. Irradiation-induced damage testing was performed on CdSe/ZnS, LaF$_3$:Eu, LaF$_3$:Ce, ZnO, and PbI$_2$ nanocrystals synthesized on a Schlenk line using appropriate solvents and precursors. Optical degradation of the nanocrystals was evaluated based on the measured dependence of their photoluminescence intensity on the irradiation dose. Radiation hardness varies significantly between various nanocrystalline material systems.

INTRODUCTION

Colloidal nanocrystals (CNCs) have attracted tremendous interest over the last few years for a wide range of applications. So far, however, their potential has generally eluded the nuclear detector community. Yet, compared to currently used scintillating particles of the micrometer size or large-size single crystals, NCs offer the prospect of significantly improved performance. Due to three-dimensional confinement and much better overlap of electron and hole wavefunctions, the optical transitions are expected to be more efficient and much faster than in bulk scintillators, which should eliminate the major problem of relatively slow response of scintillator detectors. So far, only scintillation of commercial CdSe/ZnS core/shell quantum dots under α [1,2] and γ-ray irradiation [3] has been reported, with no attempt to assess possible degradation effects of long-term exposure.

CNCs are semiconductor single crystals of ~3-100 nm size, synthesized by chemical processes. Due to their small size, the crystals exhibit enhanced quantum mechanical effects, such as size-dependent emission, room-temperature (RT) excitonic features, decrease in carrier lifetime, and dominance of surface effects. CNCs are being researched for a wide range of biomedical applications [4,5], to produce fast efficient phosphors for light emitting diodes [6,7], and as active elements in photovoltaic devices [8]. Detection of nuclear radiation by its conversion to UV or visible light can be another attractive application of CNCs.

Practical importance of a scintillator material is determined by considering characteristics such as efficiency of converting a γ photon to UV/visible light photons, conversion linearity, self-absorption of scintillation light, pulse response and decay time, energy resolution, cost, *etc.* [9,10]. One parameter, important for long-term and high radiation level measurements, is the material stability under large absorbed dose. Absorption of high-energy photons can cause multiple ionization of atoms through a combination of K shell electron ionization and ejection of multiple Auger electrons. This can form nonradiative defects within a crystal, or color centers absorbing the UV/visible light [11]. While studies of radiation hardness have been performed for numerous bulk materials [12,13], no such data are available for nanoscale scintillation materials.

For future applications of CNCs as scintillating materials, it is important to know the levels of irradiation that would degrade their optical properties. In this paper, we report the results of, to our best knowledge, the very first study of the effects of γ irradiation on photoluminescent properties of several types of CNCs.

EXPERIMENT

Optical degradation of CNCs was evaluated based on the measured dependence of their RT photoluminescence (PL) intensity on the irradiation dose. An Eberline 1000B multiple source gamma calibrator was used to study the effects of irradiation. In order to accelerate the degradation process, the strongest of the available sources was used, namely a ^{137}Cs source emitting 662 keV γ rays. The exposure rate was initially maintained at 97.3 R/h, and was later increased to 330.3 R/h when it became evident that most CNCs can sustain much higher exposures.

Irradiation experiments were performed on CdSe/ZnS, ZnO, PbI₂, LaF₃:Ce, and LaF₃:Eu CNCs synthesized on a Schlenk line using appropriate solvents and precursors. Synthesis of CdSe/ZnS core/shell CNCs, the only type of CNCs that has been investigated so far for scintillation under γ and α irradiation [1-3], was based on a modified protocol of Clapp et al. [14]. Synthesis of ZnO CNCs, of interest for optoelectronic devices and biomedical applications, was performed using a modified method of Demir et al. [15]. PbI₂ is the material that has generated interest as a scintillator [16] due to its high density and attenuation constant. PbI₂ CNCs were synthesized by using, with some modifications, the procedure of Finlayson and Sazio [17]. Lanthanum halide crystals for nuclear detection have aroused considerable interest lately [18-22]. Our synthesis of LaF₃:Ce and LaF₃:Eu CNCs was based on a modified procedure of Wang et al. [23].

Using a Horiba Jobin Yvon Fluorolog-3 spectrofluorometer, PL measurements were performed after regular weekly periods of irradiation to check if the CNCs exhibited any signs of degradation in their optical characteristics. In order to exclude the effect of natural degradation due, for example, to possible oxidative processes, each irradiated sample had its control counterpart from the same synthesis batch that was not subjected to irradiation. Assuming that both irradiated and control samples stored at RT underwent the same aging process and reacted to environmental changes in the same way, we corrected the results of PL degradation measurement of irradiated samples for any changes in PL intensity of corresponding control samples with respect to their baseline measurements prior to irradiation.

PL lifetime measurements were taken on the same spectrofluorometer in a different configuration, allowing for time-correlated single photon counting. A variety of nanosecond and picosecond pulsed LED and diode laser excitation sources were used for these experiments, depending on the CNC material and required excitation wavelength.

DISCUSSION

Fig. 1a summarizes the results of our ongoing irradiation experiments. For all the samples, the PL intensity was taken at the strongest peak of PL emission. CdSe/ZnS CNCs, initially very bright, showed a rapid loss of light output when exposed to 662 keV γ rays, and were removed from the experiment early after ~133.2 kR cumulative exposure. An examination of PL spectra of these CNCs revealed an irradiation-induced 10 nm blue shift of the PL emission peak (Fig.

1b). Characterized by a relatively low light output, the LaF$_3$:Eu and ZnO CNCs did not undergo any significant loss of PL due to the exposure. PbI$_2$ and LaF$_3$:Ce CNCs demonstrated both high levels of luminescence and stability of PL under large cumulative exposures.

Absorbed dose for different colloidal nanocrystals

The information in Fig. 1 is useful for comparative studies of the tested CNCs, but the exposure has to be converted to an absorbed dose in rads. Converting roentgens for a monoenergetic source into rads for a particular material x can be done using mass energy-absorption coefficients

$$D = 0.88E \, [\mu_{en}(h\nu)/\rho]_x/[\mu_{en}(h\nu)/\rho]_{air} \, , \qquad (1)$$

where D is the dose in rads, E is the exposure in roentgens, $h\nu$ is the γ photon energy, and $[\mu_{en}(h\nu)/\rho]_x$ is the mass energy-absorption coefficient for the subscripted material X [24].

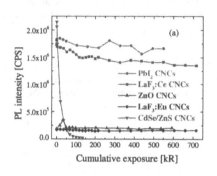 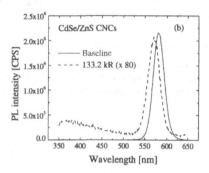

Figure 1. (a) Peak PL intensity in counts per second for selected CNCs as a function of cumulative exposure in kR. The time elapsed between data points is one week of exposure, and the increase in the distance between data points is due to an increased exposure rate. (b) Shift in PL spectrum of CdSe/ZnS CNC caused by γ-ray irradiation.

While the energy absorption coefficient for air at 662 keV is known to be 0.0293 cm^2/g [25], for CNCs its values need to be estimated using the data for constituent elements A, B, C,... [26]:

$$[\mu_{en}(h\nu)/\rho]_x = [\mu_{tr}(h\nu)/\rho]_A(1 - f_A g_A - f_B g_B - \dots)f_A + [\mu_{tr}(h\nu)/\rho]_B(1 - f_A g_A - f_B g_B - \dots)f_B + \dots . \quad (2)$$

Here, $[\mu_{tr}(h\nu)/\rho]_Y$ (Y = A, B, C...) is the mass energy-transfer coefficient for the element Y, f_Y is the weight fraction, and g_Y is the average fraction of secondary-electron energy that is lost in radiative interactions [26].

Table I. Absorbed dose conversion factors for selected CNCs at 662 keV.

CNC material	$\mu_{en}(h\nu)/\rho$ [cm^2/g]	rad/roentgen
ZnO	0.02844	0.8543
CdSe	0.02992	0.8988
LaF$_3$:Ce 5%	0.03365	1.0109
LaF$_3$:Eu 5%	0.03381	1.0157
PbI$_2$	0.04529	1.3604

With the data from Table 1 and Fig. 1, the radiation hardness of the CNCs under investigation can be compared with that of other known scintillation materials. For example, thallium doped halides, such as NaI and CsI, degrade at low absorbed doses, with detectable losses in luminescence at 1 krad, while the most radiation resistant inorganic crystal scintillator is GSO, which shows little detectable damage up to absorbed doses of 1 Grad [27].

The CNCs under test showed a wide range of radiation hardness. The CdSe/ZnS CNCs turned out to be most sensitive, having lost 50% of their light output after ~11.5 krad of absorbed dose. The other four materials demonstrated much better radiation hardness. The LaF$_3$:Ce CNCs lost 20.5% of their light output after 723.4 krad of absorbed dose. Changing the dopant to europium increased the radiation hardness, with only 9.1% of light output having been lost in LaF$_3$:Eu CNCs after 726.8 krad of dose. The absolute luminescence of LaF$_3$:Eu CNCs, however, is an order of magnitude lower than that of the LaF$_3$:Ce CNCs. The ZnO CNCs displayed no degradation in light output up to an absorbed dose of 505.9 krad. This excellent radiation hardness is compromised, however, by a very low light output of ZnO CNCs, comparable to that of LaF$_3$:Eu CNCs. Little radiation damage was observed in the PbI$_2$ CNCs which preserved a high level of luminescence up to 747.7 krad of absorbed dose, with a 9.1% decrease in PL intensity.

PL lifetime measurements

In general, the PL lifetime in CNCs is shorter than that of bulk materials, which provides advantages in positron emission tomography [28] and in radiological measurements using anticoincidence counting, where Compton radiation events can be effectively removed [29].

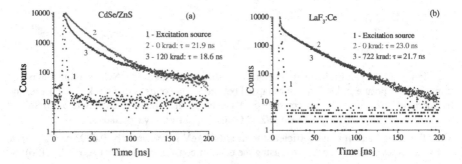

Figure 2. Effects of γ irradiation on PL lifetime for (a) CdSe/ZnS and (b) LaF$_3$:Ce CNCs. Note that there is practically no change in lifetime of LaF$_3$:Ce samples.

A PL lifetime of 21.9 ns was measured for the control CdSe/ZnS CNCs (Fig. 2a), reducing to 18.6 ns after 119.8 krad of absorbed dose, with accompanying reduction of quantum yield from 23.4% to 0.2%. The same tendency of the PL lifetime becoming shorter after irradiation was observed in LaF$_3$:Ce CNCs (Fig. 2b), although to a much lesser extent, consistent with less significant radiation damage induced in these CNCs. A PL lifetime of 23.0 ns was measured for the control LaF$_3$:Ce CNCs, twice as fast as that reported for the bulk material (42.6ns) [18]. A shorter lifetime of 21.7 ns was measured for the irradiated sample after 723.4 krad of dose. We explain these observations by the influence of nonradiative recombination centers created in the irradiated materials. Assuming the intrinsic radiative lifetimes remained unchanged, the nonradiative lifetimes associated with irradiation-induced damage can be estimated as 123.4 ns for CdSe/ZnS CNCs and 383.9 ns for LaF$_3$:Ce CNCs.

The PL lifetime of LaF_3:Eu nanoparticles synthesized by a different method was reported to be in the ms range [30]. Our measurements indicate a much shorter lifetime of 1.2 μs, which could be caused by competition with nonradiative recombination channels in uncoated CNCs. A very short PL lifetime of 100 ps was reported for bulk ZnO [31], which was beyond the resolution of our instrumentation. An instrument-limited PL lifetime of ~1 ns was observed for band-to-band transition in our ZnO CNCs. A PL lifetime of only 4.0 ns was observed in PbI_2 CNCs, which indicates their potential as very fast scintillators. No comparison could be made with bulk material, since bulk PbI_2 does not even emit light at RT [16].

CONCLUSIONS

Colloidal nanocrystals of various material systems have been tested for the effects of 662 keV γ irradiation from a ^{137}Cs source, revealing a broad range of radiation hardness. CdSe/ZnS CNCs, so far the only NC material that was used to demonstrate scintillation under α and γ irradiation, turned out to be the least radiation resistant. Changes in photoluminescence lifetime were observed, associated with the radiation damage taken by CNCs. The best combination of high PL output, good radiation hardness, and very fast PL decay was observed in PbI_2 CNCs.

ACKNOWLEDGMENTS

This work was supported by the NSF Grants IIS-0610201, CBET-0736241, and DGE-0549500. The authors are grateful to members of the UNM Department of Safety and Risk Services: Jim De Zetter, Marybeth Marcinkovich, Ralph M. Becker, Marj Walters, and Tom Rolland for their assistance with the use of Eberline 1000B multi-source gamma calibrator. Canberra Albuquerque, Inc. is gratefully acknowledged for the loan of a calibrated Canberra Radiac Meter Geiger-Müller counter, which was used to calibrate the Eberline ^{137}Cs source used in our tests.

REFERENCES

1. S. Dai, S. Saengkerdsub, H.-J. Im, A. C. Stephan, S. M. Mahurin, "Nanocrystal-based scintillators for radiation detection", Unattended Radiation Sensor Systems for Remote Applications, 15-17 April 2002, Washington, DC, AIP Conf. Proc. **632**, pp. 220-224, 2002.
2. S. E. Létant, T.-F. Wang, "Study of porous glass doped with quantum dots or laser dyes under alpha irradiation", Appl. Phys. Lett. **88** (10), Art. 103110, 8 March 2006.
3. S. E. Létant, T. F. Wang, "Semiconductor quantum dot scintillation under γ-ray irradiation", Nano Lett. **6** (12), 2877-2880, 13 Dec. 2006.
4. M. Osiński, K. Yamamoto, T. M. Jovin (Eds.), *Colloidal Quantum Dots for Biomedical Applications*, Proc. SPIE **6096** (2006).
5. M. Osiński, T. M. Jovin, K. Yamamoto (Eds.), *Colloidal Quantum Dots for Biomedical Applications II*, Proc. SPIE **6448** (2007).
6. I. Matsui, "Nanoparticles for electronic device applications: A brief review", J. Chem. Eng. Japan **38** (8), 535-546, Aug. 2005.
7. Y.-Q. Li, A. Rizzo, R. Cingolani, G. Gigli, "White-light-emitting diodes using semiconductor nanocrystals", Microchimica Acta **159** (3-4), 207-215, July 2007.
8. V. I. Klimov, "Mechanisms for photogeneration and recombination of multiexcitons in semiconductor nanocrystals: Implications for lasing and solar energy conversion", *J. Phys. Chem. B*, **110** (#34), pp. 16827-16845, 31 Aug. 2006.
9. G. F. Knoll, *Radiation Detection and Measurement* (3rd Ed.), John Wiley & Sons, New York 2000, p. 219.
10. M. Nikl, "Scintillation detectors for x-rays", Meas. Sci. Technol. **17** (4), R37-R54 (2006).

11. M. Nikl, P. Bohacek, E. Mihokova, S. Baccaro, A. Vedda, M. Diemoz, E. Longo, M. Kobayashi, E. Auffray, P. Lecoq, "Radiation damage processes in wide-gap scintillating crystals. New scintillation materials", Nucl. Phys. B (Proc. Suppl.) 78, 471-478 (1999).

12. S. Normand, A. Iltis, F. Bernard, T. Domenech, and P. Delacour, "Resistance to gamma irradiation of LaBr$_3$:Ce and LaCl$_3$:Ce single crystals", Nucl. Instrum. Methods Phys. Res. Sect. A 572 (2), 754-759, 11 March 2007

13. W. Drozdowski, P. Dorenbos, A. J. J. Bos, S. Kraft, E. J. Buis, E. Maddox, A. Owens, F. G. A. Quarati, C. Dathy, V. Ouspenski, "Gamma-ray induced radiation damage in LaBr$_3$:5%Ce and LaCl$_3$:10%Ce scintillators", IEEE Trans. Nucl. Sci. 54 (#4, Pt. 3), 1387-1391, Aug. 2007.

14. A. R. Clapp, E. R. Goldman, H. Mattoussi, "Capping of CdSe–ZnS quantum dots with DHLA and subsequent conjugation with proteins", Nat. Protocols 1 (3), 1258-1266 (2006).

15. M. M. Demir, R. Muñoz-Espi, I. Lieberwirth, G. Wegner, "Precipitation of monodisperse ZnO nanocrystals via acid-catalyzed esterification of zinc acetate", J. Mater. Chem. 16 (28), 2940-2947 (2006).

16. M. K. Klintenberg, M. J. Weber, D. E. Derenzo, "Luminescence and scintillation of PbI$_2$ and HgI$_2$", J. Luminescence 102-103, 287-290, May 2003.

17. C. E. Finlayson, P. J. A. Sazio, "Highly efficient blue photoluminescence from colloidal lead-iodide nanoparticles", J. Phys. D: Appl. Phys. 39 (8), 1477-1480, 21 April 2006.

18. A. J. Wojtowicz, M. Balcerzyk, A. Lempicki, "Optical spectroscopy and scintillation mechanisms of Ce$_x$La$_{1-x}$F$_3$", Phys. Rev. B 49 (21), 14880-14895, 1 June 1994.

19. J. Glodo, W. W. Moses, W. M. Higgens, E. V. D. van Loef, P. Wong, S. E. Derenzo, M. J. Weber, K. S. Shah, "Effects of Ce Concentration on Scintillation Properties of LaBr$_3$:Ce", IEEE Trans. Nucl. Sci. 52 (5), 1805-1808, Oct. 2005.

20. G. Bizarri, J. T. M. de Haas, P. Dorenbos, C. W. E. van Eijk, "Scintillation properties of Ø 1×1 inch^3 LaBr$_3$: 5%Ce^{3+} crystal", IEEE Trans. Nucl. Sci. 53 (2), 615-619, April 2006.

21. K. W. Krämer, P. Dorenbos, H. U. Güdel, C. W. E. van Eijk, "Development and characterization of highly efficient new cerium doped rare earth halide scintillator materials", J. Mater. Chem. 16 (27), 2773-2780 (2006).

22. G. Bizarri, P. Dorenbos, "Charge carrier and exciton dynamics in LaBr$_3$:Ce^{3+} scintillators: Experiment and model", Phys. Rev. B 75 (18), Art. 184302, May 2007.

23. F. Wang, Y. Zhang, X.-P. Fan, M.-Q. Wang, "One-pot synthesis of chitosan/LaF$_3$:Eu^{3+} nanocrystals for bio-applications", Nanotechnology 17 (5), 1527-1532, 14 March 2006.

24. P. W. Cattaneo, "Calibration procedure for irradiation tests on silicon devices", IEEE Trans. Nucl. Sci. 38 (3), 894-900 (1991).

25. K. Eckerman, Radiological Toolbox Computer Program, Oak Ridge National Lab., 2003.

26. F. H. Attix, Introduction to Radiological Physics and Radiation Dosimetry, John Wiley & Sons, New York 1986, pp.155-157.

27. G. F. Knoll, op. cit., pp. 243, 247.

28. G. Muehllehner, J. S Karp, "Positron emission tomography", Phys. Med. Biol. 51 (13), pp. R117-R137, 7 July 2006.

29. G. F. Knoll, op. cit., pp.771-772.

30. J. W. Stouwdam, F. C. van Veggel, "Improvement in the luminescence properties and processability of LaF$_3$/Ln and LaPO$_4$/Ln nanoparticles by surface modification", Langmuir 20 (26), 11763-11771, 21 Dec. 2004.

31. J. Wilkinson, K. B. Ucer, R. T. Willams, "Picosecond excitonic luminescence in ZnO and other wide-gap semiconductors", Radiation Measurements 38, 501-505 (2004).

AUTHOR INDEX

SUBJECT INDEX

Printed in the United States
By Bookmasters